The Great Survivors
220 million years of turtle evolution

Justin Gerlach

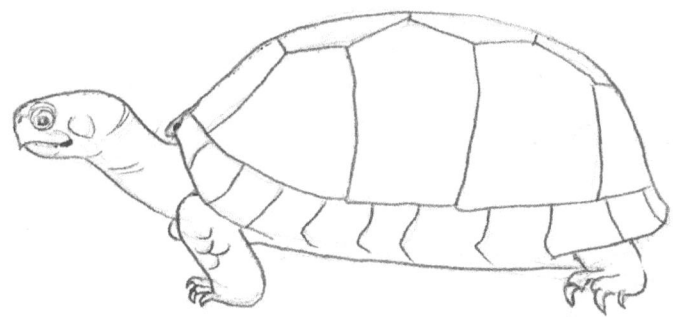

ISBN: 978-0-9533787-5-3

Phelsuma Press,
Cambridge, U.K.

The Great Survivors
220 million years of turtle evolution

1. Introduction

With their bizarre, primitive appearance, tortoises and turtles are among the strangest of the animals alive today. Not only do they look remarkable they have changed relatively little over many millions of years - recognisable turtle fossils are older than even the dinosaurs.

The most obvious defining feature of the order Chelonia is the shell, made up of a bony dorsal carapace and a ventral plastron, both covered with horny plates (scutes). The skeleton is also unusual in association with the shell: the back-bone is often fused to the carapace; the shoulder-girdle is particularly highly modified, being within the rib cage and with the scapula and coracoid bones being at least partly fused together. The acromion process on the scapula joins on to the plastron, forming a y-shaped shoulder. They have many other distinctive features that are either unique or shared with few other animals. These relatively obscure aspects of the limbs and the skull are particularly useful when trying to identify fragmentary fossils: the skull usually has a curved quadrate bone surrounding the tympanic membrane of the ear; the squamosal bone is small relative to the quadrate and quadratojugal bones; there are no post-parietal or postfrontal bones; and they are toothless.

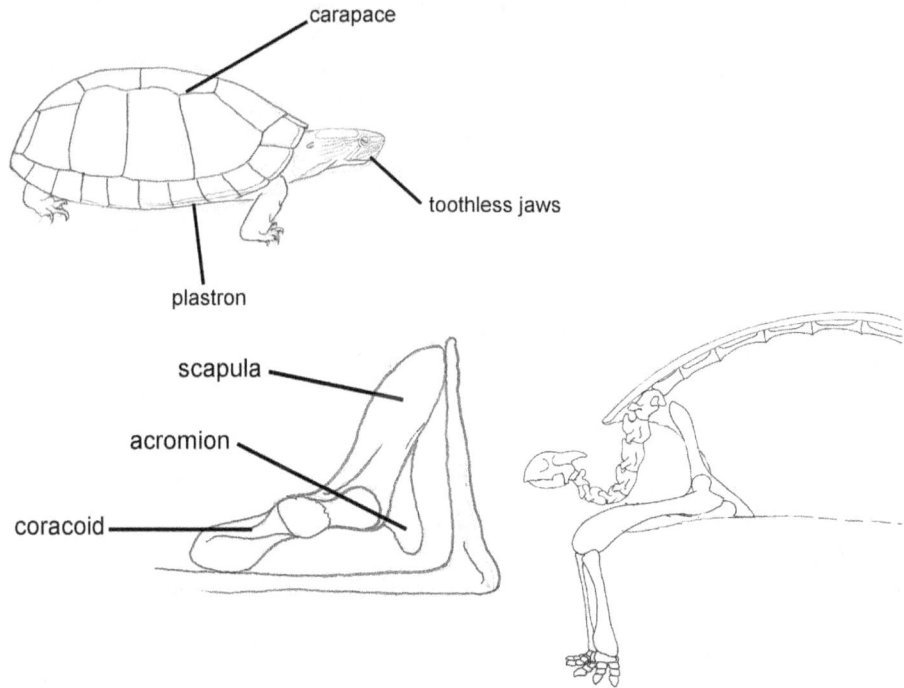

Features of turtles: external features, shoulder girdle and fore-limb in association with the shell showing the butressing of the scapula agains the shell and the coracoid and acromion process against the plastron.

Their origins of turtles have long been debated; they look primitive but are so different from other living vertebrates that it is difficult to say what their closest relatives are. With new techniques such as DNA analysis we are discovering that what little we thought we were certain of is incorrect; old and apparently implausible ideas need to be re-examined. Not all of the old ideas need to be given close attention, for example George-Louis Leclerc, Comte de Buffon (1707-1787) considered the turtle to be a 'crustaceous fish', similar to lobsters, an idea that can now easily be disregarded.

In 1757 Carl von Linnée (often known as Linnaeus) published the 10th edition of '*Systemae Naturae*', the work from which the binomial naming system is deemed to have originated. In this he placed turtles in the reptiles and distinguished 11 forms in a single genus *Testudo*. These were *Mydas*, (the green turtle *Chelonia mydas*), *Caretta* (the loggerhead *Caretta caretta*), *orbicularis* and *lutaria* (European pond turtle *Emys orbicularis*), *scabra* (possibly the spot legged turtle *Rhinoclemmys punctularia*), *graeca* and *pusilla* (spur-thighed tortoise *Testudo graeca*), *carolina* and *carinata* (Carolina box turtle *Terrapene carolina*), *geometrica* (geometric tortoise *Psammobates geometricus*) and *serpentina* (snapping turtle *Chelydra serpentina*). These represented turtles that were familiar in Europe (the marine turtles, the pond turtle and the Mediterranean tortoises) or had been brought to Europe as curiosities from North America. These 11 species in the reptiles was an improvement on the first edition in 1735 which listed four species which he considered to be amphibians. By his last edition (12th) in 1766 the list had grown to 15 species.

Numerous collectors in the 18th and early 19th century expanded the range of turtle species available to the developing taxonomic community. Notable 18th century works on reptiles that included turtles were published by Schneider in 1783 (20 species), Count de Lacepède in 1788 (23 species), Gmelin in 1789 (33 species) Schoepff in 1792 (54 species) and Daudin in 1801 (58 species). In 1812 August Schweigger published his '*Prodromus Monographiae Cheloniorum*', listing 78 species based on specimens he had studied in European museums, largely in Paris. These species were now divided into the genera *Chelonia, Chelydra, Chelys, Emys, Testudo* and *Trionyx*. In 1820 Merrem reduced this to 62 species in four genera. The same number of species were used by Wagler in 1830 but he pushed the number of genera up to 21. A great increase in recognised diversity occurred in 1834 with the publication of '*Erpétologie générale ou Histoire naturelle complète des reptiles*' by Duméril and Bibron who considered 121 species to be divided into 22 genera. In 1826 Fitzinger published the '*Neue Classification der Reptilen*' with 122 species in 10 genera and 37 subgenera.

The most prolific early worker on turtles was Gray who published numerous papers and monographs from 1831 onwards. In his final compilation in 1873 he recorded 187 species in 93 genera. Gray's work was followed and revised by Boulenger in his '*Catalogue of the Chelonians, Rhynchocephalians, and Crocodiles in the British Museum (Natural History)*' in 1889 with 201 species. Further increases occurred in the next few years with Siebenrock recording 232 species in 57 genera in 1909. A classic work for much of the 20th century was '*Die rezenten Schildkröten, Krokodile und Brückenechsen*' by Mertens and Wermuth in 1955. This cut the number of recognised species down to 211 but increased the number of genera to 66. The number of species

6

has continued to increase steadily through several publications from the late 20th century, with David in 1994 recording 273 species in 90 genera.

The 21st century has seen two major attempts to review the complex and confused array of turtle names. In 2007 Fritz and Havas published a review aimed at clarifying species names for the regulation of international trade (CITES), concluding that there were 313 species in 97 genera. This list was superceded by the Turtles of the World checklist of the Turtle Taxonomy Working Group. The latest version of this list (2010) records 328 species and a further 124 subspecies.

Distinguishing turtle species has been remarkably contentious, as indicated by the presence of 1203 different names for the 452 currently recognised subspecies and species of turtle. This has been due to many different factors, particularly the variability of many characters; the frequent transport of the animals and the resultant confusion over their origins; and confusion over the interpretation of genetic data. Originally different species were distinguished on the basis of the shell shape, scute arrangement and the colour pattern. Whilst some species are easily distinguished this way, some very distantly related species can look very similar despite a very distinctive looking shell. This is seen in the radiating pattern of the geometric tortoise *Psammobates geometricus*, the radiated tortoise *Astrochelys radiata* and the Indian starred tortoise *Geochelone elegans*. By 1835 skeletal characters were also being considered. Much of the variation in skeleton relates to function, such as having limb bones that reflect being aquatic, semi-aquatic, semi-terrestrial or fully terrestrial. The skull has been more informative, as long as the differences in feeding behaviour are taken into account. This use of skeleton and shell features has allowed most species to be reasonably well defined; however, considerable confusion remains with species groups that have undergone notable radiations, such as some of the river turtles, the Mediterranean tortoises and some giant tortoises. For these the distinctions between populations and subspecies, between subspecies and species, and where exactly the boundaries lie remain purely matters of opinion.

In recent years there has been an increasing emphasis on defining species by using genetic techniques. These complement morphological studies and may provide useful insights into the significance of differences between populations. Such molecular studies started with comparison of protein structures which provided a general estimate of genetic differences between samples. These can be very effective in this aim but provide little specific data on exactly what the differences are. Comparing the numbers of chromosomes between species has been useful in some animal groups, such as mammals. In turtles it may have some use in the pleurodire side-necked turtle group where usually 50-60 chromosomes are present (except for 28 in the Podocnemididae and 34-36 in the Pelomedusidae), but in other turtles there is very little variation in number. In soft-shelled turtles there are always 66 (or 68 in the Carettochelyininae). Most others have 56 (sea turtles, Dermatemydidae, Kinosternidae [except for Stuarotpyinae with 54) or 52 (geoemydids [except *Siebenrockeilla* 50] and tortoises [except for 54 in one gopher tortoise]). The lowest numbers are found in the Emydidae terrapins which usually have 50, but range from 52 down to 48. With little sensible pattern, chromsome analysis has not be of any use in clarifying turtle relationships.

7

The first turtle phylogeny, as proposed by Hay in 1908.

8

More precise studies using specific gene sequences have largely replaced proteins. Genes are sequenced from the nucleus of the cell or from mitochondria within cells. Studies of nuclear DNA can be carried out with a focus on specific genes or on 'microsatellite DNA'. Microsatellites are highly repeated segments of DNA that mutate quite easily, these can be used to shed light on population structure and gene flow over a relatively short time period, although the interpretations can be tricky and the mathematical assumptions underlying the analysis are very confusing and highly contentious. These studies are undoubtedly useful in clarifying evolutionary relationships but do have some significant problems when it comes to delimiting species boundaries. Mitochondria are small organelles within cells that carry out much of the cell's energy transfer processes. Billions of years ago these were separate, free-living bacteria that have become integrated with our cells. They retain a small amount of DNA which is passed down the mitochondrial line, separate from the cell's nuclear DNA. The mitochondria are supposedly inherited only in the egg and thus should provide data on maternal inheritance, unlike the nuclear DNA which is inherited from both parents. However, there is increasing evidence of at least occasional paternal inheritance as well which may confuse some analyses. Most of the alternative analysis techniques have yet to be used in turtles, so genetic results still have to be considered preliminary in most, if not all, species of turtle.

Another aspect of genetic analysis is the 'molecular clock'. This uses the amount of genetic difference between animals (individuals, species, genera or any other level of comparison) and rates of mutation to calculate the time at which the animals diverged, or when they last shared a common ancestor. This assumes that mutations are accumulated in the DNA randomly in a clock-like manner; it is now abundantly clear that mutation rates are not actually constant and that any clock-like behaviour is very approximate. If the mutation rate can be checked at different points in the study it is possible to tell whether or not the molecular clock behaves in a useful way in this group of organisms. Determining the mautation rate requires that at least one accurately dated point is known, this provides a fixed point from which calculations of rates can be made. This is the big problem for turtles – we have very few reliably dated points in the turtle fossil record, and as a result few molecular clock calculations for turtles can be considered reliable. Most studies use a 'standard' rate for some genes, which comes from a rough estimate made in one study. In the following account of turtle evolution I have used many of these molecular clock calculations where they seem to make evolutionary and geological sense, but these dates should not be assumed to be accurate; they are merely indications. The same caveat applies to the dating of the fossils themselves as the geological record of turtles is riddled with fossils for which only approximate or contradictory dates are available.

Although tortoises play an important role in the development of evolutionary thought through the role of the Galapagos giant tortoises in Charles Darwin's recognition of evolution, the evolution of turtles themselves was not discussed until Hay in 1908 published an evolutionary tree of living and fossil families. Even after this first step no significant evolutionary studies were carried out until late in the second half of the 20th century with the exception of the rather superficial analysis made by Loveridge and Williams in 1952. Matters changed in 1972 when Gaffney started publishing his detailed studies of fossil turtles, providing evolutionary hypotheses for the features and species discovered.

Understanding of the evolution of turtles included various skeletal features and the study of changes in the pattern of arteries in the head. Skeletal studies have included analysis of the obvious features of the position, size and shape of bones in the skull and in the shell, and also more obscure aspects such as the arrangement of bones in the feet. The study of cranial circulation identified a large foramen stapedio-temporale for a large stapedial artery in tortoises and terrapins which have a smaller internal carotid artery than the soft-shelled turtles, with a vestigial palatine artery. The soft-shells and their close relatives have a smaller stapedial artery and within the group there is a large palatine artery in musk turtles and a large mandibular artery in soft-shells. These seem to have some evolutionary relevance but are at least in part associated with feeding specialisations as the large arteries provide an enhanced blood supply to the major jaw muscles used in different types of feeding. If different animals evolve to live in a similar manner, in this case feeding in a very similar way, they are likely to adapt their anatomy along similar lines and may end up resembling one another closely. Such convergence is a major problem in evolutionary studies - recognising the evolutionarily meaningful characters from amongst the convergences is something of and art, and is not surprisingly often contentious. This is particularly problematic when analysing fossils, particularly when the fossils are very fragmentary.

The fossil record ought to be of great value in evolutionary studies and turtles make particularly good fossils with their solid bony shells. There is an abundance of fossil shell material but as with the shells of the living species, this is not particularly informative. The more useful skull material is relatively scarce. Marine and freshwater turtles live in environments that provide good fossils and consequently the fossil records of these are reasonably well known. For the freshwater species there is an abundance of material and some good studies have been made on many of these, although almost every new find radically changes the story. In contrast the terrestrial tortoises often live in arid environments that are very poor for fossilisation; good tortoise material is very

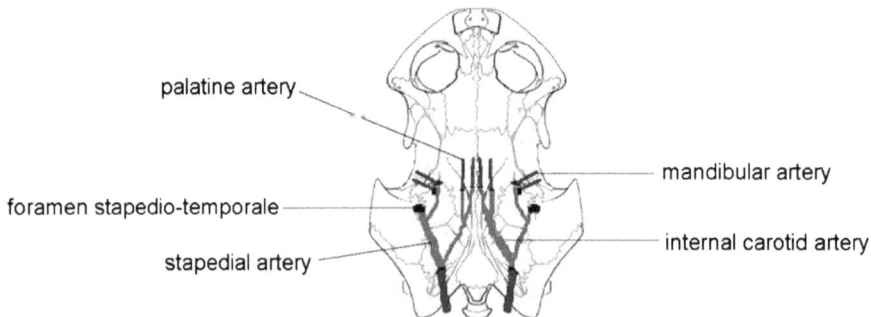

Cranial circulation in turtles. Left side shows a pattern of an enlarged stapedial arterly, right side shows an enlarged internal coarotid artery.

scarce. Interpretation of the fossils was attempted in 1952 by Loveridge and Williams, but this ended up placing most remains in the largely meaningless groups of '*Testudo*' (small tortoises) and '*Geochelone*' (large tortoises). Some local tortoise faunas have been well studied, particularly the North American gopher tortoises *Gopherus*. For the early turtles the fossil record is patchier and there is still considerable uncertainty over their relationships. This uncertainty covers an enormous time period; the earliest fossils (from China) are currently estimated to be 220 million years old, but no clear picture of turtle evolution emerges until 100 million years ago at the earliest. These dates are highly approximate, being based on dating using calculations based on the decay of radioactive elements found in fossils. The accuracy of this dating varies with the quality of the sample and its age; the oldest turtle remains could be dated anywhere within a 15 million year period (220-235 million years). The following chapters discuss the origins of turtles and their early diversification during that confused period.

References

Adler, K. 2007. The development of systematic reviews of the turtles of the world. *Vert. Zool.* **57**: 139-148

Fritz, U. & P. Havas. 2007. Checklist of Chelonians of the World. *Vert. Zool.* **57**: 149-368

Iverson J.B., R.M. Brown, T.S. Akre, T.J. Near, M. Le, R.C. Thomson & D.E. Starkey. 2007. Search for the Tree of Life for Turtles. *Chel. Res. Monogr.* **4**: 85-106

Jamniczky, H.A. & A.P. Russell. 2007. Reappraisal of patterns of nonmarine cryptodiran turtle carotid circulation: evidence from osteological correlates and soft tissues. *J. Morphol.* **268**: 571-587

Linnaeus, C. 1758. *Systema Naturae.* 10th ed.

Loveridge, A. & E.E. Williams. 1957. Revision of the African Tortoises and Turtles of the Suborder Cryptodira. *Bull. Mus. Comp. Zool.* **115**(6): 163-557

Turtle Taxonomy Working Group [Rhodin, A.G.J., P.P. van Dijk, J.B. Iverson & H.B. Shaffer]. 2010. Turtles of the world, 2010 update: annotated checklist of taxonomy, synonymy, distribution, and conservation status. In: Rhodin, A.G.J., P.C.H. Pritchard, P.P. van Dijk, R.A. Saumure, K.A. Buhlmann, J.B. Iverson & R.A. Mittermeier (Eds.). Conservation Biology of Freshwater Turtles and Tortoises: A Compilation Project of the IUCN/SSC Tortoise and Freshwater Turtle Specialist Group. *Chel. Res. Monogr.* **5**: 000.85-000.164

2. The origin of turtles

Since palaeontology and evolutionary biology became established studies over the past 100 years, turtles have generally been considered to be the only surviving lineage of primitive reptiles. This view comes from their rather primitive appearance and, in particular, the structure of their skulls. Conventionally terrestrial vertebrates are divided into three major groupings based on nature of openings in the side of the skull as first defined by Williston in 1917. The 'fenestrae' (windows) define the groupings of the anapsids (turtles and many fossil vertebrates), diapsids (birds, crocodiles, lizards, snakes and many fossils) and synapsids (mammals and their ancestors). Anapsids have solid skulls, with no openings on the sides except for those for the nostrils, ears and eyes; diapsids have an additional two distinct openings; and synapsids have two openings which have fused into one large opening (Fig. 1). The anapsid nature of the turtle skull is not always clearly recognisable as almost all living turtle groups have a tendency to open out the side of the skull, not through the development of windows but through 'temporal emargination'. In this the back margin of the skull moves forwards, narrowing the bones of the cheek region, and in some cases leading to their loss entirely.

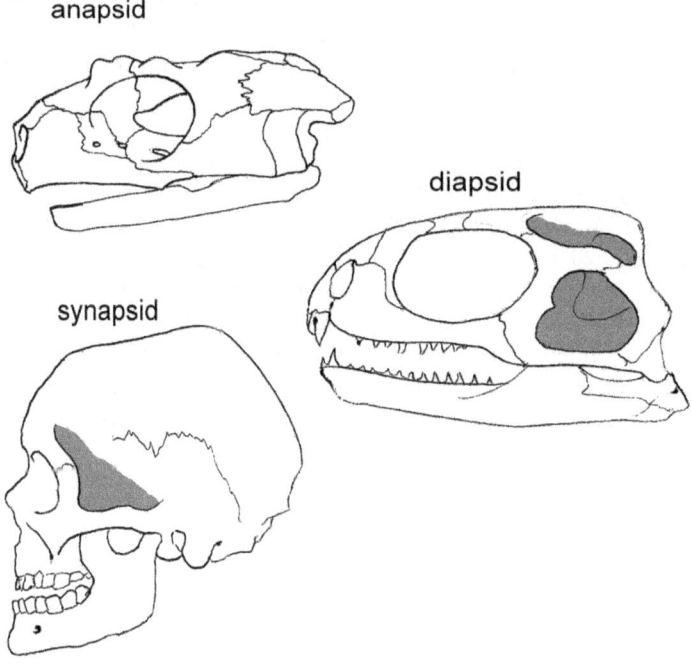

Fenestration patterns, fenestrae shaded.

This arrangement places the turtles (as anapsids) near the base of the family tree of the living groups, retaining the primitive skull form. This has led to many attempts to identify turtle relatives in the fossil anapsids. Early suggestions were that turtles may have been descended from the captorhinomorps, in particular species of the genus *Seymouria*. Most of the similarities seem rather superficial and a more convincing case was made for turtles being descended from pareiasaurs, most recently by Lee in 1993.

Pareiasaurs were large reptiles, between 60 cm and 3 metres in length. These were all heavily built, and often armoured, herbivores of the Permian era. Armour consisted of bony plates in the skin (osteoderms) of varying degrees of development, from isolated bony discs to interlocking plates and spines. Their skulls were very robust, with projecting knobs and sometimes horns. The teeth were distinctively leaf-shaped, resembling those of iguanas. In limb structure, armouring and in the robust skull they resemble turtles, but these could all be adaptations to their large, heavy, herbivorous life-style. The basic-looking skull lacks an emargination at the back (associated with the tympanic membrane), but this may be a specialisation increasing the robustness of the skull rather than a primitive absence. The skull had several notable features such as a ventral flange on the quadratojugal bone, a posterior extension of the squamosal, covering the normal quadrate emargination, a boss on the supratemporal and a ventral process on the angular of the lower jaw. These features are not found in other animals, including turtles.

Pareiasaurs and ancestral turtles have similar arrangements of the bones of the palate, with the internal nostrils near the centre of the mouth (far from the jaw edge). This is in contrast to mammals where the internal nostrils are right at the back of the mouth and most reptiles where they are near the front. The back of the skull is strongly reinforced, with the bones around the ear region being ossified (separating the inner ear from the brain) and some bones fused together. The basisphenoid and basioccipital bones at the base of the skull are fused, therefore lacking the ventral 'otic fissure' that usually separates them. These features suggest either a shared ancestry or similar patterns of stresses on the skull resulting from similar diets and ways of feeding. The rest of the skeleton has some odd similarities such as all of the tail vertebrae having prominent lateral projections (only on the first five tail vertebrae in most early reptiles). More importantly the acromion process on the scapula looks similar to that of turtles, giveing the shoulder-blade a very distinctive form seen only in turtles

Pareiasaurus

in living vertebrates. This feature is probably associated with the movement constraints of having a sprawling gait, a relatively heavy body and very little body flexibility. This would also explain why both groups are similar in other limb features; both have a distinct pit on the humerus (the 'ectepicondylar foramen') for strong muscle attachment (this is also present in some synapsids and diapsids); the main process of the femur (its greater trochanter) for muscle attachment is on the dorsal side; the fifth toe is reduced and there is a prominent supporting ridge over the top of the hip socket in the pelvis. Most of these similarities could be explained as convergences of rather rigid bodies.

Of all the parieasaur *Sclerosaurus* is the most turtle-like, sharing with turtles the reduction in the number of back vertebrate (also in the aberrant pelycosaur *Eunotosaurus*); a tall, narrow scapula with a similarly shaped pit (glenoid fossa); reduced digits (also in some caseid pelycosaurs); fusion of the astragalus and calcaneum in the ankle; dermal armour on the back (also in the seymouriamorph *Kotlassia*); and loss of ventral ribs or gastralia. The dwarf genera *Anthodon*, *Nanoparia* and *Pumiliopareia* share with turtles the dermal armouring, flattened ribs, cylindrical scapula and similar limb bone structures. Again, the similarities between *Sclerosaurus*, dwarf pareiasaurs and turtles are mainly associated with moving a heavy, semi-rigid body. A study of the microscopic structure of the dermal armour showed that pareiasaur and turtle armour has little in common, pariesaur armour having ornamental bosses, different patterns of blood vessels, and a uniform internal structure, whereas turtle armour has distinct inner and outer bone layers. This suggests that the armour developed in very different ways in the two groups, making a close relationship highly unlikely.

A particular problem in understanding turtle evolution has been interpreting the scapulocoracoid, the fusion of the scapula and coracoid bones. This shoulder-girdle is within the rib cage of turtles instead of on the outside of the skeleton. Drawing affinities between turtles and other groups needs to explain how the scapulocoracoid moved under the ribs. In proposing a parieasaur origin for turtles Lee followed an earlier suggestion (made by Watson in 1914) that a broad, flat carapace formed from laterally flared ribs and dermal armour evolved behind the scapulocoracoid which then migrated backwards under the rib cage. Parieasaur ribs were held laterally and slightly curved, and the shoulder girdle was narrower than the massive rib cage, enabling, at least hypothetically, a migration under the rib cage during development. This seems to solve the main problems in turtle evolution quite neatly – the massive herbivorous pareiasaurs developed solid skulls in association with their diet, reducing their teeth in favour of a constantly growing beak and developed dermal armour, ultimately fusing the rib cage and armour into a carapace which encloses the shoulder girdle. This would make the turtles the last survivors of the ancient, primitive anapsid reptiles.

However, in the last few years DNA studies of the relationships of living vertebrates have tended to place the turtles within the diapsids. Initially this was viewed as a mistake but the same result keeps cropping up in different studies and it is worth questioning whether the divisions of anapsid, diapsid and synapsid are as meaningful as has been assumed. Back in 1918 the great herpetologist Boulenger observed that the closed temporal region of the anapsid turtle skull does not have quite the same pattern of bones as in the extinct early reptiles, suggesting that it might be secondarily closed.

14

He considered that a fenestrated pelvis (an opening between the pubic and ischial bones, or 'thyroid fenestra') and a hooked fifth metatarsal were more significant and supported grouping turtles with diapsids such as the tuatara *Sphenodon*.

It now seems certain that turtles are indeed diapsids and are most closely related to crocodiles and birds. Although it goes against recent orthodoxy this idea is not really new, having been proposed by Goodrich in 1916. More recently (in 1996) it was revived by Rieppel and de Braga. Goodrich noted the similarity of turtle and diapsid fifth metatarsals and aortic arches. This means that the anapsid skull of the turtle is far from being primitive; they must have evolved from a properly diapsid ancestor and reverted to what looks like the primitive condition. It seems that turtles are part of a group of very strange reptile descendants, which includes the living crocodiles and birds and the extinct dinosaurs. At present we cannot place them any more precisely, which leaves open several questions and reopens others – such as where do the pariaeasaurs really fit in?

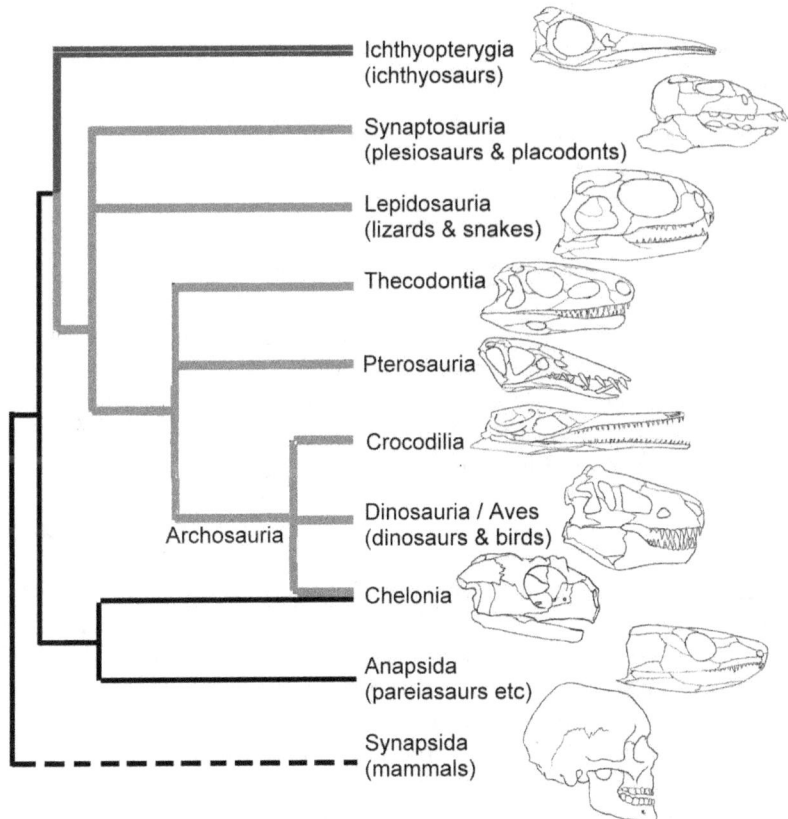

Phylogeny of anapsids (black), diapsids (grey, with the slightly different 'euryapsids' in dark grey) and synapsids (dashed), Chelonia shown in the traditional anapsid relationship and the more probable diapsid relationship.

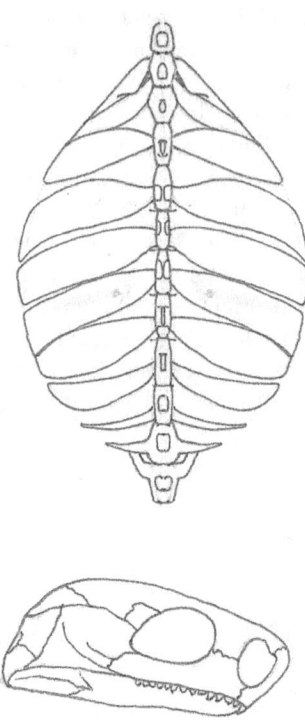

Eunotosaurus skeleton showing the expanded ribs, and skull.

One of the oldest fossils suggested to be a possible turtle ancestor is *Eunotosaurus africanus* from the Permian of South Africa. This 260 million year old fossil was suggested to be an ancestor of the turtles on the basis of its extremely broad ribs which resemble possible precursors of the turtle shell. In addition it had turtle-like vertebrae but possessed teeth and had a normal shoulder girdle construction and the skull differed from all turtles in possessing an epipterygoid bone (replaced by a process of the parietal in turtles). The anapsid skull is only very superficially like that of turtles and this species is now thought to be a 'parareptile' totally unrelated to turtles.

Comparisons of parareptiles with turtles have relied on *Proganochelys quenstedtii* as the model early turtle. *Proganochelys* is a large, terrestrial animal and this suggested that turtles evolved on land. For over 120 years this remained the oldest known turtle, but in 2008 this honour fell to a very different animal. The oldest turtle currently known is *Odontochelys semistestacea* which was discovered in China. This species is a bizarre animal in that it had a turtle's plastron but lacked a carapace. The 220 million year old fossil had a fully developed plastron, identical to that of modern turtles, but on the back it only had a series of bony plates running down the spine and very broad ribs, the ribs themselves lacking bony plates. Unlike modern turtles it still retained teeth and had a relatively long skull and tail. The presence of a plastron but no carapace suggests that it was defended against damage (or attack) from below and it is

16

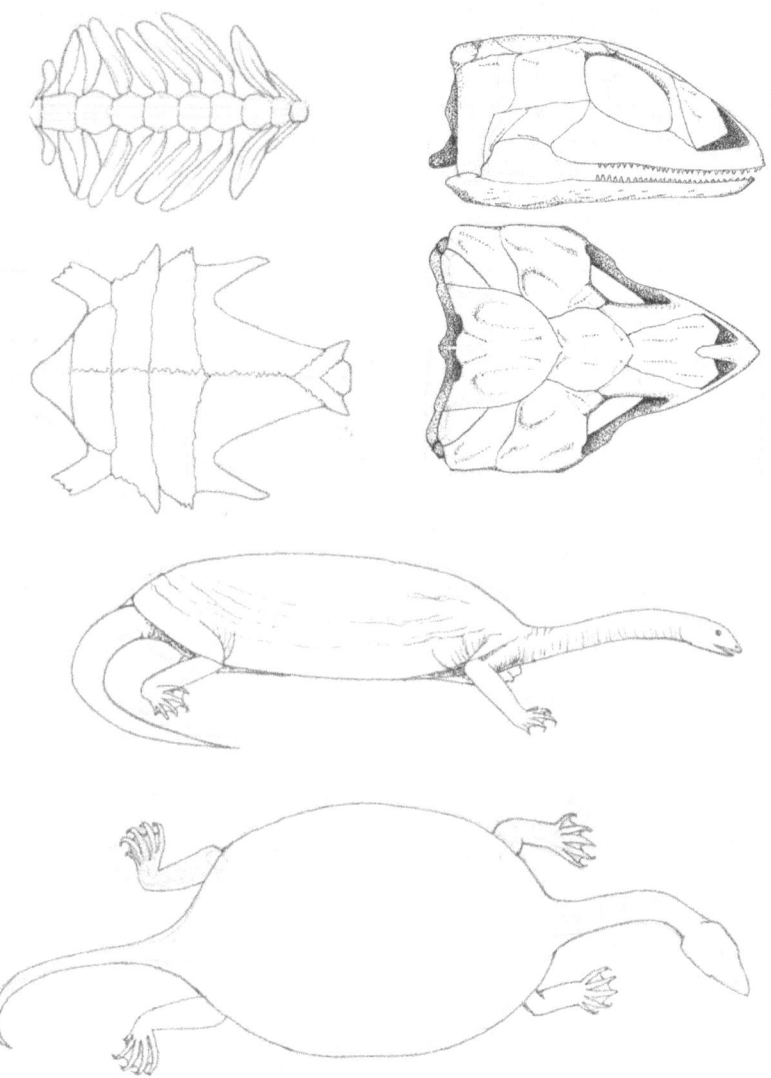

Odontochelys carapace bones, plastron, skull and reconstruction.

thought to have been an aquatic species. Being aquatic seems a likely original life-style for all turtles, all of which, whether living or fossil, have forward-facing nostrils which are typical of aquatic vertebrates, rather than the lateral nostrils of terrestrial reptiles. These details are important as turtles were originally thought to have evolved in water and colonised land relatively late, but *Proganochelys* is clearly a terrestrial animal. The discovery of *Proganochelys* therefore changed the picture and was often interpreted to show that turtles were originally terrestrial but then returned to the water

17

for much of their evolution. Now *Odontochelys* reverses the picture again: turtles were originally aquatic after all, but the turtle body-plan is so well adapted to life on land that the terrestrial environment was colonised very quickly and repeatedly. Turtles have probably changed from aquatic to terrestrial and back to aquatic over and over again.

Although *Proganochelys* of Germany and Thailand lived at about the same time as *Odontochelys* (around 220 million years ago) it now seems to be a less primitive turtle. This splendidly preserved species had the classic turtle features of a shell, with internally positioned girdles. It had eight cervical and 10 dorsal vertebrae, a pleurodire neck articulation (see Chapter 4) and a pelvis fused to the carapace (subsequently lost in cryptodires) but retained proper teeth on the roof of the mouth, although not on the jaws. The forelimbs of *Proganochelys* are essentially like those of modern turtles. It differs obviously from modern turtles in having a long tail covered in heavy bony spikes and spikes on the neck. The head does not seem to have been capable of retraction into the shell and the armoury of the tail suggests that it was used defensively on land (similarly to the tail club of ankylosaur dinosaurs). They were about a metre long. The skull of *Proganochelys* retained some characters typical of many other vertebrates, but lost by more recent turtles: the presence of lachrymal, supratemporal, postfrontal and postparietal bones, two vomers (instead of a single one) and the separation of the braincase from the skull roof. The modern turtle skull has a reduction in the number of

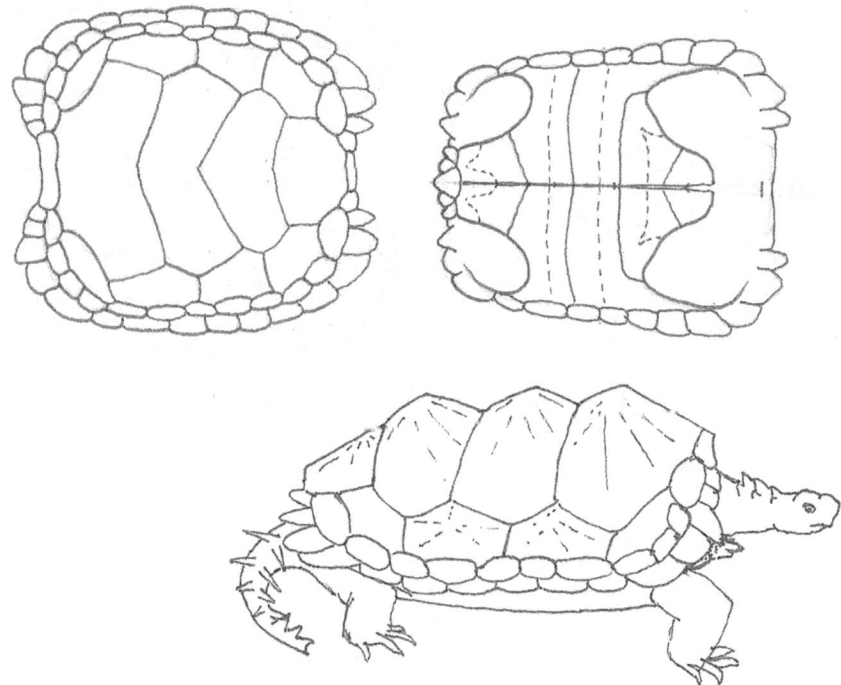

Proganochelys shell, skull and reconstruction.

of bones in the skull, fusion of the vomers and fusion of the braincase with the skull and an associated strong contact between the opisthotic, quadrate and squamosal bones. These features appear 185 million years ago in the more modern turtles, the Casichelydia, such as *Proterochersis robusta* and *Kayentachelys aprix*.

Over the past 20 years we have developed a reasonably good understanding of the appearance of the earliest turtles such as *Odontochelys* and *Proganochelys*, but there is still a big gap between some form of basic lizard-like diapsid reptile and even a partially shelled turtle: where does the shell come from?

References

Boulenger, G.A. 1918. Sur la place des chéloniens dans la classification. *Comp. Rend. Acad.émie Des Sci., Paris* **169**: 605– 607

Gaffney, E.S. 1990. The comparative osteology of the Triassic turtle *Proganochelys*. *Bull. AMNH* **194**

Hedges, S.B., & L.L. Poling. 1999. A molecular phylogeny of reptiles. *Science* **83**: 998–1001

Iwabe, N., Y. Hara, Y. Kumazawa, K. Shibamoto, Y. Saito, T. Miyata & K. Katoh 2004. Sister Group Relationship of Turtles to the Bird-Crocodilian Clade Revealed by Nuclear DNA–Coded Proteins. *Mol. Biol. Evol.* **22**

Kirsch, J.A.W. & G.C. Mayer 1998. The platypus is not a rodent: DNA hybridization, amniote phylogeny and the palimpsest theory. *Phil. Trans. R. Soc. Lond. B* **353**: 1221-1237

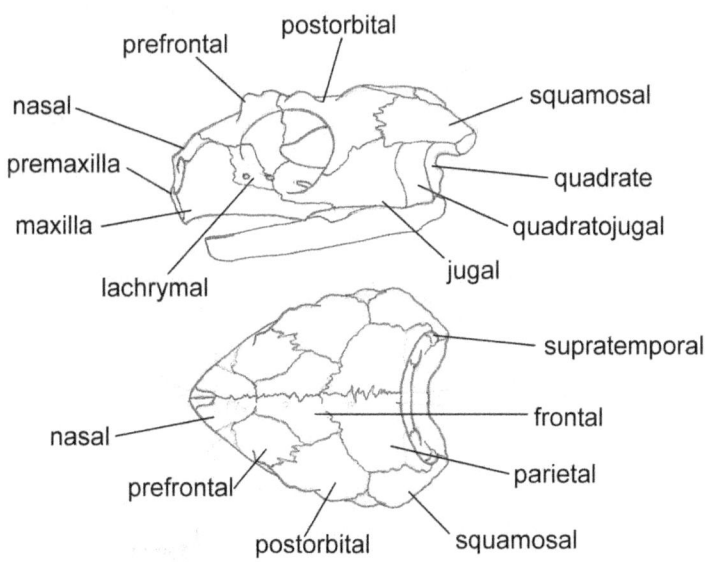

Proganochelys skull.

Kumazawa, Y. & M. Nishida. 1999. Complete mitochondrial DNA sequences of the green turtle and blue-tailed mole skink, statistical evidence for archosaurian affinity of turtles. *Mol. Biol. Evol.* **16**: 784–792

Lee, M.S.Y. 1993. The origin of the turtle body plan: bridging a famous morphological gap. *Science* **261**: 1716-1720

Lee, M.S.Y. 1997. Pareiasaur phylogeny and the origin of turtles. *Zool. J. Linn. Soc.* **120**(3): 197-280

Lee, M.S.Y.1998. Similarity, parsimony and conjectures of homology: The chelonian shoulder girdle revisited. *Journ. Evol. Biol.* **11**: 379

Li, C., X.-C. Wu, O. Rieppel, L.-T. Wang & L.J. Zhao. 2008. An ancestral turtle from the Late Triassic of southwestern China. *Nature* **456**(7221): 497-501

Rieppel, O. 2000. Turtles as diapsid reptiles. *Zool. Script.* **29**: 199–212.

Rieppel, O. & R.R. Reisz. 1999. The origin and early evolution of turtles. *Ann. Rev. Ecol. Syst.* **30**: 1-22

Schyer, T.M. & P.M. Sander. 2009. Bone microstructures and mode of skeletogenesis in osteoderms of three pareiasaur taxa from the Permian of South Africa. *J. Evol Biol .* **22**: 1153–1162

Sues, H.-D. & R.R. Reisz. 2008. Anatomy and Phylogenetic Relationships of *Sclerosaurus armatus* (Amniota: Parareptilia) from the Buntsandstein (Triassic) of Europe. *Journ. Vert. Palaeont.* **28**(4):1031-1042.

Valenzuela, V. & D.C. Adams. 2011. Chromosome number and sex determination coevolve in turtles. *Evolution* **65**(6): 1808–1813

Zardoya, R. & A. Meyer 1998. Complete mitochondrial genome suggests diapsid affinities of turtles. *Proc. Natl. Acad. Sci. USA* **95**: 14226-14231

Zardoya, R. & A. Meyer. 2001. The evolutionary position of turtles revised. *Naturwiss.* **88**: 193-200

3. The origin of the turtle shell

If, as seems probable, turtles are very specialised diapsids, and not living pareiasaurs, how should we interpret the shell? The pareiasaur link seemed to solve most problems, with different species indicating different stages of rib expansion, dermal plate development, fusion of ribs and armour and ultimately the movement of the shoulder-girdle. If the ancestor of the turtles was not some sort of pareiasaur but a diapsid, how did its shell evolve? We can only solve this problem by understanding exactly what the turtle shell is and how it forms.

The turtle carapace comprises broad, flat ribs fused to neural and costal bones in the dermis. These bones have long been assumed to originate in the outer layers of the skin (dermal bones), as first proposed by the German physiologist Carl Gustav Carus in 1834. They have been assumed to originate as osteoderms found in such early land vertebrates as pareiasaurs. More recent embryological studies have, however, determined that the ribs themselves stimulate ossification of the costal bones, which makes it unlikely that the shell is the result of the fusion of two separate structures (ribs and osteoderms). Instead, it is now thought that as the ribs start to ossify they secrete

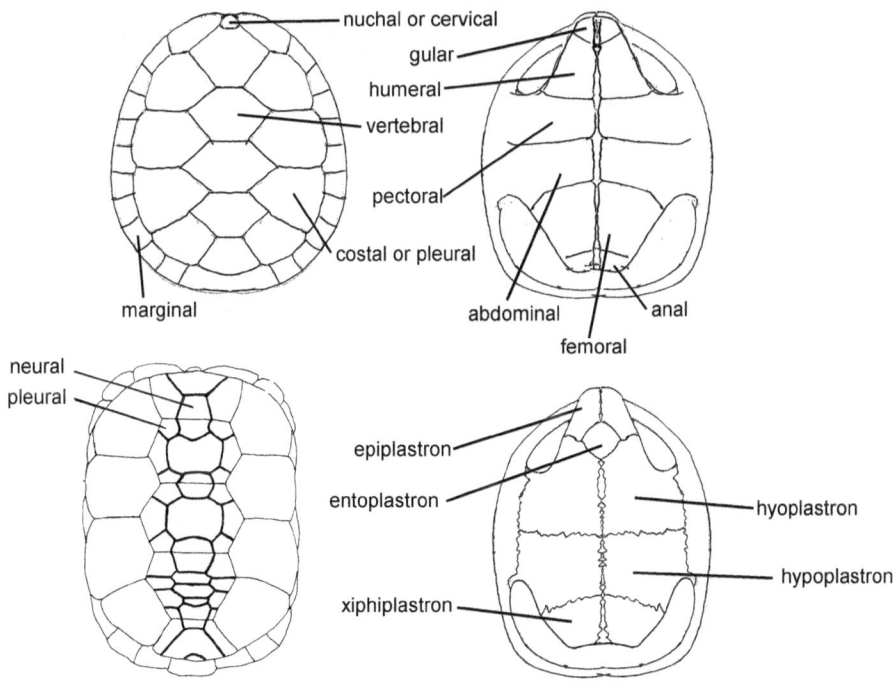

Anatomy of the turtle shell - carapace scutes (top left), plastral scutes (top right), carapace bones (bottom left), plastral bones (bottom right.

21

proteins that stimulate bone development in the membranes overlying the ribs. The 9th to 18th vertebrae and their associated ribs grow laterally into the outer layers of the body wall. The corresponding ventral ribs which normally form the complete rib-cage do not develop in turtles and are replaced by the plastron. This completed bony covering of carapace and plastron encloses the shoulder girdle.

An important part of the puzzle of the origin of the turtle shell was solved in 2005 when it was found that the turtle shell forms from a ridge along either side of the embryo (the carapacial ridge); cells in this ridge secrete a protein called fibroblast growth factor which attracts the cells that will form the ribs into the carapacial ridge. In the absence of this growth factor the rib cells migrate to the underside of the body wall and in the absence of the genes for the carapacial ridge the ribs develop but do not expand. Once the ribs are in the dermis they undergo normal endochondral ossification (cartilage turning into bone) and in turn signal for the dermal ossification (skin layers becoming bone); thus the ribs trigger the development of the dermal shell plates.

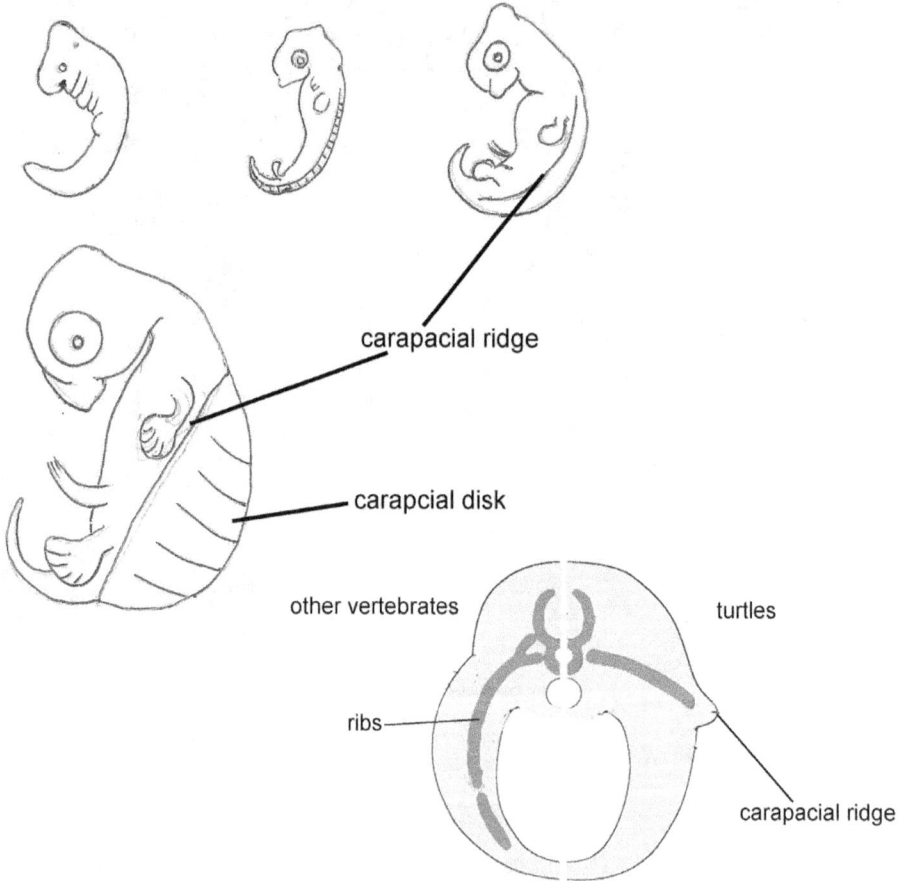

Turtle embryonic development, showing the development of the carapacial ridge, and cross section of an embryo showing the ribs being drawn towards the carapacial ridge.

The carapacial ridge is controlled by specific genes that are found in other vertebrates, but with a different function; it appears that the first stage in the evolution of the turtle shell was not the evolution of new genes but an alteration in the function of existing genes. The lateral expansion of the ribs causes the anterior ribs to move over the scapula, enclosing it in the rib cage (and consequently in the carapace). The ribs form the main part of the carapace, but there are also several additional separate bone-forming regions at the front (nuchal) and back (pygal). The nuchal is the only part of the carapace that really is dermal bone, whilst the pygal seems to be a separate, later addition. The origin of the turtle shell seems to be a combination of a new structure (the carapacial ridge) and a process of developing existing tissues (ribs and bony plates) in a new location (in the dermis rather than in the musculature). Altering the cell signalling pathways would allow the appearance of turtles in the fossil record with little in the way of obvious ancestral forms. The complete carapace develops rapidly from a chain of developmental steps triggered by a small number of genes that have altered from their basic vertebrate function. In contrast the plastron forms by intramembranous ossification in nine distinct ossification centres. As these centres expand they interact, leading to interdigitation of the plastral bones. These bones are dermal in origin and may be derived from the protective bones of the belly ('gastralia') seen in crocodiles and alligators, and in many dinosaur fossils. Whilst the gastralia are technically not dermal, but rather bones forming within membranes, their development is triggered in the same way.

There appear to be three main stages in the evolution of the turtle body-plan: development of the carapacial ridge and the expanded ribs, dermal ossification and plastron formation. These could all evolve independently, although there is an obvious link between the ribs and the carapace. The earliest fossil turtles could be expected to have both the broadened ribs and a plastron (such as *Odontochelys*) or just one of these characteristics. This would probably be quickly followed by the full turtle structure.

Most distinctive features of turtles relate to their possession of a shell, and this has naturally been the main subject of research on turtle evolution. The more obscure aspects of turtle biology can also be interpreted in the light of recent research. For example, the pattern of sex determination in turtles is now known to be linked to processes which are much more widespread than was originally thought. Although the exact pattern of temperature dependent sex determination (TSD) differs between turtles and crocodilians (females at high temperatures in turtles, at low temperatures in alligators and at intermediate temperatures in crocodiles) the underlying process is the same: a gene that apparently triggers male development is activated at specific temperatures. The difference between turtles and crocodilians lies simply in the exact temperature of activation, which is a result of differences in protein structure. It turns out that this male determining gene is present in all vertebrates, and in both males and females. It is not the presence of the gene that makes males (unlike the male determining SRY gene on the Y chromosome of mammals) but the activity of the gene. Thus the effects of the gene can be influenced by the environment (by temperature increasing or decreasing its activity) or by dosage (two copies of this same gene on the Z chromosome make birds with 'ZZ' chromosomes male, whilst one makes them female – 'ZW'). In some species the two processes can interact – temperature can stimulate or inhibit gene activity, effectively overriding the genetic sex of an individual. Even mammals have this gene, although in us it has been supplanted by the X and Y chromosomes and is only present in an inactive, degraded form. In turtles the temperature system has changed to a genetic system several times. This has been linked to changes in chromosome number (groups that have acquired genetic sex determination seem also to have a high frequency of changes in chromosomes) and these seem to have occurred at times when global temperatures increased rapidly. At the moment these are just correlations and quite what they mean is obscure. Genetic sex determination is known from *Glyptemys insculpta, Pangashura smithii, Siebenrockeilla crassicollis*, the Staurotypinae, the Trionychinae and Chelidae. The mechanisms of genetic sex determination are still largely unknown, having only been examined in very few species. A ZZ/ZW system is known in *Pangashura smithii* and *Pelodiscus sinensis* which may function like the ZW system in birds. An XX/XY system is known in *Acanthochelys radiolata, Stuarotypus slavinii, S. triporcatus, Siebernockiella crassicollis, Chelodina longicollis* and *Emydura maquarii*, exactly how this works is not known, if these turtles still use the dosage gene it may be that their so-called Y chromosome is derived from a Z chromosome with an extra copy of the gene

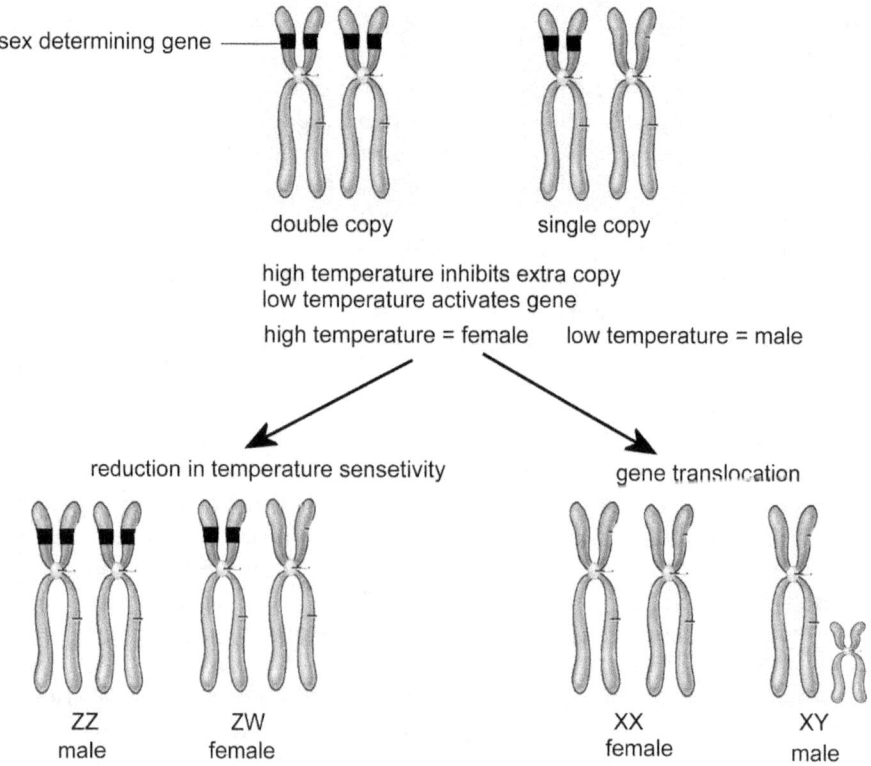

Possible routes of the evolution of different sex determination mechanisms.

With the new knowledge of the development of the turtle shell and the recent fossils finds we can reconstruct the evolutionary history of the earliest turtles. It is probable that their ancestor was a diapsid reptile, related to the living crocodiles. This would be an aquatic or semi-aquatic animal, protected by a tough skin, containing osteoderms, just like a crocodile. Its skeleton would have been like that of any 'normal' diapsid, with openings into the side of the skull and with teeth. A shift towards a snapping bite, rather than the complex jaw action of most diapsids would have changed the pattern of forces in the skull, resulting in an apparent return to the anapsid condition. Coincidental with this, bony plates formed, particularly on the ventral surface, and the embryonic carapacial ridge started to evolved, giving rise to early partially shelled turtles such as *Odontochelys*. Continuation of this process would have led to the evolution of the full turtle body plan, probably still as an aquatic animal. It seems that colonisation of land occurred very early on in the turtle group, with the appearance of *Proganochelys*. Thus, by 200 million years ago several forms of turtles were present in fresh-water and some on land.

At the time of the origin of turtles some 220 million years ago, the terrestrial environment formed a single large land mass – Pangaea. Over a large part of the terrestrial environment arid conditions predominated. This time saw the earliest known of the dinosaurs with *Eoraptor* in Argentina and the earliest mammals with the small shrew-like *Adelobasileus* in North America. Other relatives of the dinosaurs, such as crocodiles, dominated the terrestrial vertebrate fauna. Primitive temnospondyl 'amphibians' were still present in Europe, North America and Australia. Turtles seem to have evolved in the cooler, wetter north-eastern regions of Pangaea. Forest habitats were dominated by ferns and some early seed plants such as ginkgoes. Pangaea started to break up around 200 million years ago. The causes of this are not known; it may have been related to an extraordinary surge in volcanic activity or a meteorite strike in Canada. Whatever the cause, the end of the Triassic period was associated

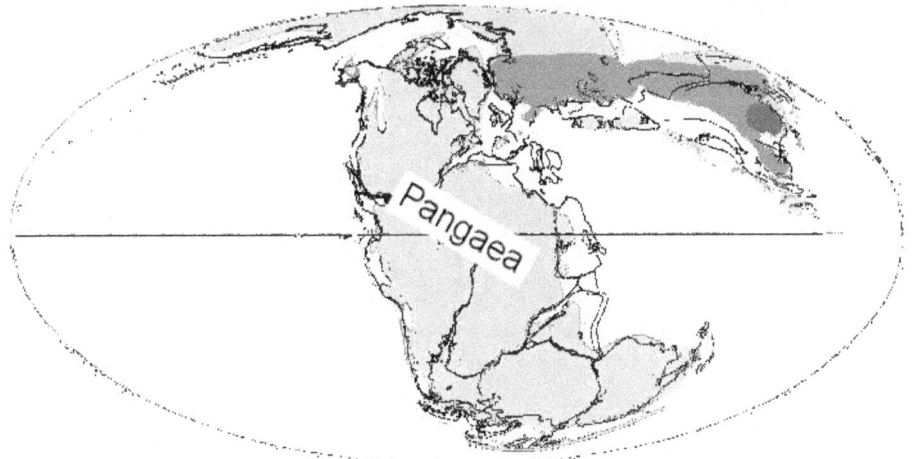

Distribution of the earliest turtles 200-220 million years ago – *Odontochelys* (dark grey) and *Proganochelys* (mid-grey).

with a mass extinction event, leaving several herbivorous archosaur groups extinct, opening the way to the dominance of the dinosaurs and the diversification of the turtles.

References

Cebra-Thomas, J., F. Tan, S. Sistla, E. Estes, G. Benderi, C. Kim, P. Riccio & S.F. Gilbert. 2005. How the turtle forms its shell: a paracrine hypothesis of carapace formation. *J. Exp. Zool. (Mol. Dev. Evol.)* **304B**: 558-569

Gilbert, S.F., G. Bender, E. Betters, M. Yin & J.A. Cebra-Thomas. 2007. The contribution of neural crest cells to the nuchal bone and plastron of the turtle shell. *Int. Comp.Biol.* **47**(3): 401-408

Gilbert, S.F., G.A. Loredo, A. Brukman & A.C. Burke. 2001. Morphogenesis of the turtle shell: the development of a novel structure in tetrapod evolution *Evol. Devel.* **3**: 47-50

Kuraku, S., R. Usuda & S. Kuratani. 2005. Comprehensive survey of carapacial ridge-specific genes in turtle implies co-option of some regulatory genes in carapace evolution. *Evol. Devel.* **7**: 3

Martinez, P.A., T. Ezaz, N. Valenzuela, A. Georges & J.A. Marshall Graves. 2008. An XX/XY heteromorphic sex chromosome system in the Australian chelid turtle *Emydura macquarii*: A new piece in the puzzle of sex chromosome evolution in turtles. *Chromosome Res.* **16**: 815–825

Nagashima, H., S. Kuraku, K. Uchida, Y. Kawashima-Ohya, Y. Narita & S. Kuratani. 2007. On the carapacial ridge in turtle embryos: its developmental origin, function and the chelonian body plan. *Development* **134**: 2219-2226

Nagashima, H., F. Sugahara, M. Takechi, R. Ericsson, Y. Kawashima-Ohya, Y. Narita & S. Kuratani. 2009 Evolution of the Turtle Body Plan by the Folding and Creation of New Muscle Connections. *Science* **325**: 193-196.

Valenzuela, V. & D.C. Adams. 2011. Chromosome number and sex determination coevolve in turtles. *Evolution* **65**(6): 1808–1813

4. Origins of the main modern turtle groups

There appear to have been a range of turtle species by 200 million years ago. A typical example of these is *Chinlechelys tenerstesta* from New Mexico. This is a very fragmentary fossil, comprising a few pieces of shell and plastron and an osteoderm. Little can be said with confidence regarding this specimen beyond noting that it was similar to *Proganochelys* in having prominent osteodermal neck spines and that the carapace dermal plates were only weakly attached to the ribs and vertebrae.

The living turtles can be divided into the side-necked turtles (pleurodires) and the 'hidden-necked' cryptodires. The pleurodires retract their necks into the shell sideways, in contrast to the vertical neck retraction of cryptodires (Fig. 1). The two groups also differ notably in their skulls, particularly in respect of how the muscles for raising the lower jaw run onto the skull. In cryptodires the lower jaw closing muscles run over a process on the otic chamber formed by the prootic and quadrate bones, the 'processus trochlearis oticum'. Pleurodires lack this process and instead the jaw closing muscles run over a flange on the pterygoid bone of the palate. As a result of this difference in muscle arrangement pleurodires tend to have a strong snapping jaw closing action, whereas cryptodires have a more powerful but slower closure. This makes pleurodires effective aquatic predators but poor herbivores in comparison with cryptodires; all pleurodires are at least partially carnivorous whereas cryptodires include carnivores, omnivores and strict herbivores.

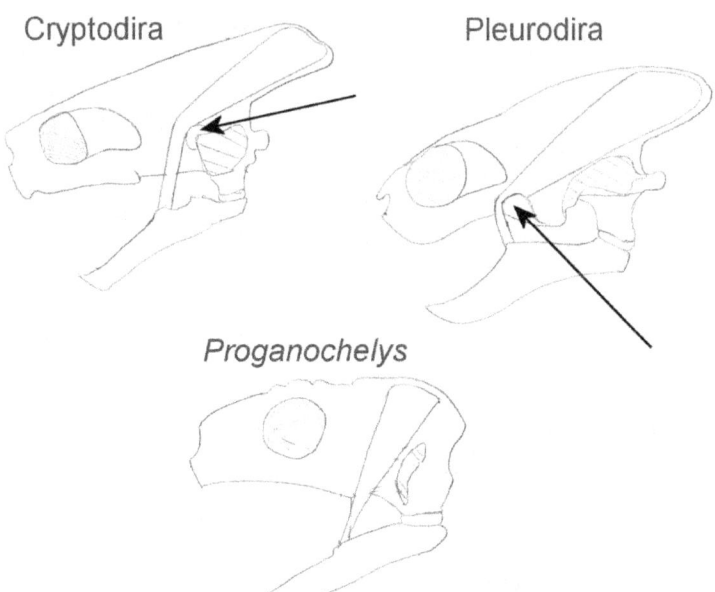

Cryptodire, pleurodire and *Proganochelys* skulls showing the processus trochlearis oticum in cryptodires and the pterygoid process in pleurodires.

In early turtles such as *Proganochelys* the jaw muscles attach in much the same place as in cryptodires, although there is no clear processus trochlearis oticum. It seems probable that the cryptodire jaw closing mechanism is the primitive condition. There are other characteristics of cryptodires and pleurodires, but these are often difficult to identify in fossils. For example, in cryptodires there is a vertical flange on the external process of the pterygoid bone; in *Proganochelys* there is a similar, but much smaller process; and in pleurodires the process is laterally directed and interacts with the jaw muscles. The cryptodires always have contact between the prefrontal and the vomer bone; in *Proganochelys* the prefrontal does not reach the vomer, nor does it in pleurodires. In cryptodires the canal for the carotid artery is entirely enclosed by the pterygoid bone, whereas in other turtles it passes through the basisphenoid or is between these two. The big differences between cryptodires and pleurodires probably result from pleurodires specialising from a generalist turtle stock, while cryptodires have retained most of the generalist characters. Early pleurodires are all flattened aquatic animals, whilst cryptodires were the more robust forms. Flattening of the pleurodire form would have led to two major changes: preventing vertical neck retraction, resulting in the modification of the vertebrae to allow horizontal retraction, and flattening and broadening the skull. An increase in the width of the skull may have caused the external process of the pterygoid bone to interfere with the jaw closing musculature, leading to the muscles running over the process and the subsequent loss of the processus trochlearis oticum. The cryptodire positioning of the carotid artery seems to result from a change in timing of bone and blood vessel development; what is usually considered the basisphenoid bone in turtles is really the parabasisphenoid, a fusion of the basisphenoid and the parasphenoid bones. If those fuse early (cryptodires) this structure pushes the carotid arteries to the sides, into the developing pterygoids. If fusion occurs later the arteries may be trapped between the basisphenoid and parasphenoid sections.

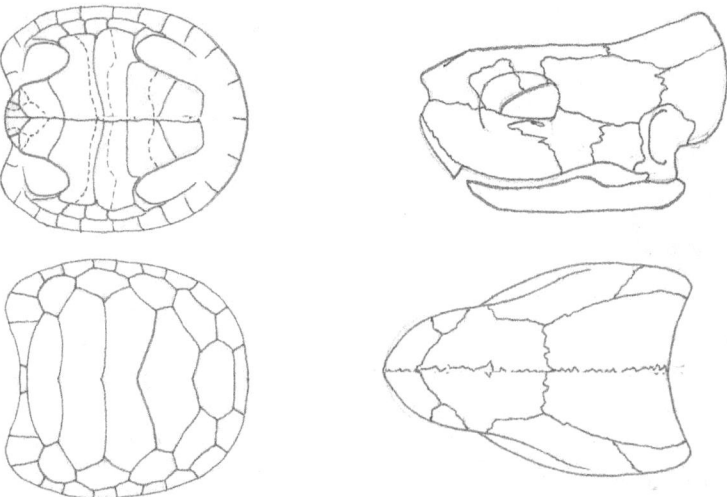

Kayentachelys plastron, carapace and skull.

Kayentachelys, an aquatic turtle from Arizona (185 million years ago) and *Proterochersis robusta* from Germany (210 million years ago) were supposed to be the earliest cryptodires and pleurodires. *Kayentachelys* was originally reported to have the cryptodire jaw closing mechanism but as described above, this is not diagnostic of cryptodires and closer examination suggests that it did not have the fully developed cryptodire features. Many of the features of this species are primitive, including the presence of teeth on the pterygoid bones. *Proterochersis* was supposed to be a pleurodire because its pelvis was fused to the carapace. This is a pleurodire character but is occasionally found in other turtles. *Kayentachelys* and *Proterochersis* are now thought to be early turtles, predating the split between cryptodires and pleurodires. *Proterochersis* was probably terrestrial but *Kayentachelys* was probably aquatic.

In early turtle evolution there is a clear process of skull bone reduction and simplification. In most terrestrial vertebrates the skull bones in the snout have the frontal bones preceded by nasal bones or prefrontal bones, or even both in some species. Modern turtles have prefrontals rather than nasals, but *Proganochelys* retained both prefrontals and nasals, as did *Kayentachelys* and the later *Pleurosternon*. One living turtle still has nasals – the matamata *Chelis fimbriatus*, which seems to retain the primitive character.

Along with these supposed cryptodires and pleurodires several early turtles were found in different parts of the world 200 million year ago. These included *Palaeochersis talampayensis* in South America (Argentina). This was a large (70 cm), apparently terrestrial species. Other turtles of the time included the aquatic *Heckerochelys romani* of Russia. Thus by 200 million years ago turtles were found throughout Pangaea. The main diversity of turtles seems to have been in the tropical and subtropical areas which provided ideal conditions for these early, largely semi-aquatic turtles.

Kayentachelys

Proterochersis

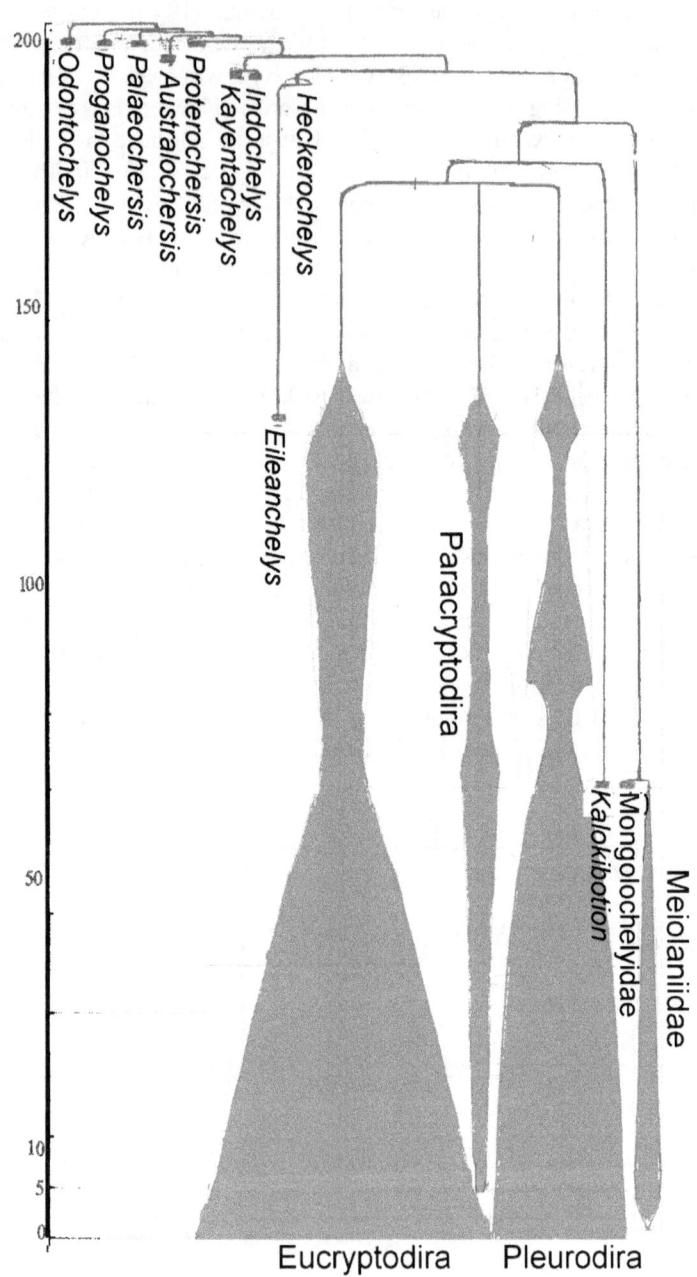

Phylogeny of early turtles, age (millions of years ago) on left.

By the time that Pangaea broke up into the northern supercontinent of Laurasia and the southern Gondwana turtles were already present on both land masses. Early Gondwanan species included the South African *Australochelys africanus* from 185 million years ago and the Indian *Indochelys spatulata* 145 million years ago. Laurasian forms were represented by *Dinochelys whitei* from North America 150 million years ago and *Eileanchelys waldmani* from the Isle of Skye, Scotland 165 million years ago. *Eileanchelys* seems to be related to the earlier Russian *Heckerochelys* and was an aquatic species from a closed water system on the margins of the continent that alternated between being a low-salinity lagoon and a freshwater floodplain lake. Whether this species was exclusively freshwater or partly marine is not known.

By 140 million years ago a wide range of turtle forms had appeared across the world. The most noteworthy animals from this stage of turtle evolution are *Mongolochelys* and the Meiolaniidae. *Mongolochelys efremovi* is known from many specimens and most of the skeleton is known. These were all found in a small area of Nemegt Svita in the western Gobi Desert of Mongolia and seem to be the remains of mass mortality events caused by droughts drying up water bodies. They were around 20-25 cm long and appear to have been rather terrapin-like. Originally described as belonging to the living family Dermatemydidae they are now thought to belong to the poorly defined early radiation, with only a superficial similarity to any living group. They are currently placed in the 'Mongolochelydiae'. *Mongolochelys* seems to be most closely related to the Australian *Otwayemys cunicularius*, although this is hard to make sense of on geographical grounds. If the relationship between them is correct these forms of turtles were presumably spread throughout the world. Both of these seem to form a sister group to the Meiolaniidae, with which *Mongolochelys* shares the presence of well defined scutes on the head and also some trace of horn-like protuberances on the skull.

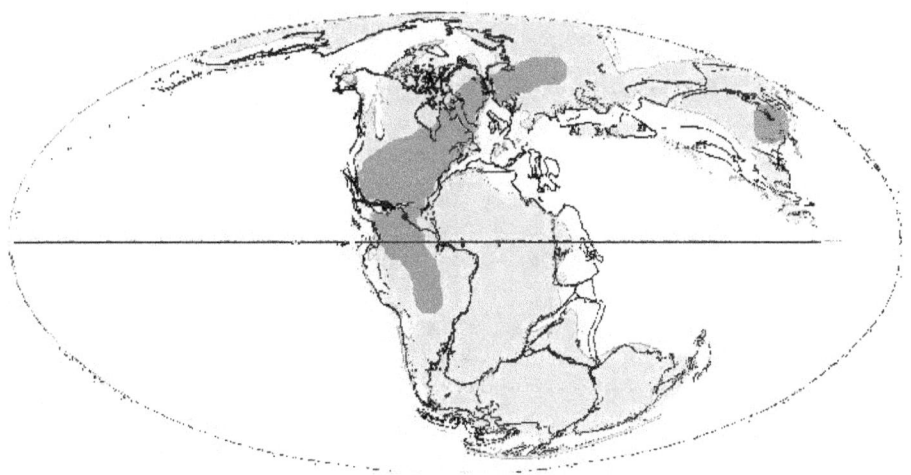

Distribution of the early turtles 200-140 million years ago.

The Meiolanidae are perhaps the most interesting of the early turtles. *Meiolania* was first known from a vertebra. This was identified as a monitor lizard by Richard Owen in 1881, and named *Meiolania* ('small butcher') to distinguish it from the larger lizard *Megalania* ('great butcher'). Six years later Thomas Henry Huxley associated it with further remains, allowing it to be identified as a turtle. Meiolaniids are diagnosed by their horns, which are formed from the squamosal and supraoccipital bones. The skull looks primitive in that there is no temporal emargination and the nasals are well developed (being comparable to those of *Proganochelys*) although the family extends back to 56 million years ago. Fossils of the genus are known from 34 million years ago to only 3,000 years ago. With their horns and large size (reaching 2.5 m in length) they were highly distinctive animals. The lateral horns gave the skull a width of up to 60 centimetres. In every way this was the largest of all land turtles. The horns would have prevented proper neck retraction, but with this armouring that would hardly have been necessary. The tail was also armoured with plates, spines and a terminal club. Reconstructions of *Meiolania* and its relatives seem to show that meiolaniids evolved in the Cretaceous landmass comprising South America, Antarctica

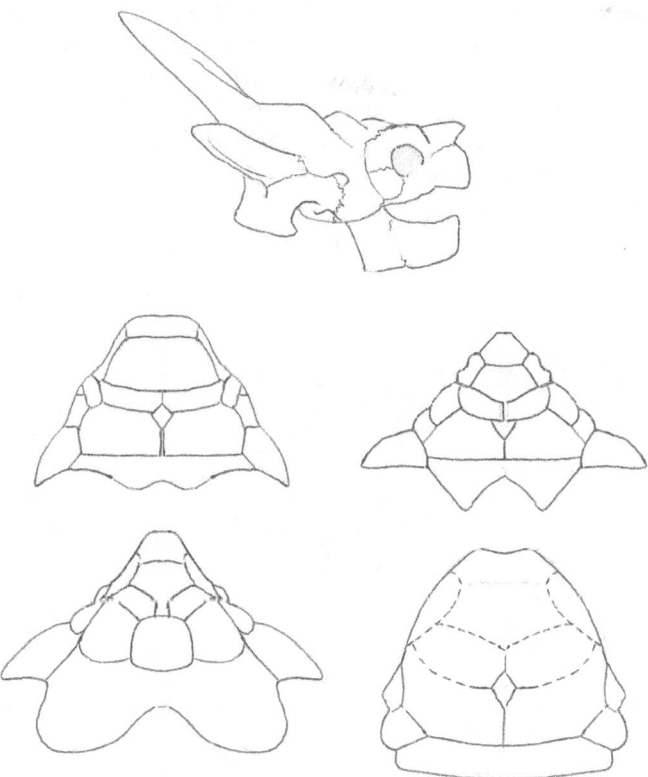

Meiolaniidae skulls in lateral and dorsal views
top: *Niolamia argentina*; middle: *Meiolania platyceps* (left) and *Ninjemys oweni* (right);
bottom: *Niolamia argentina* (left) and *Warkalania carinaminor* (right).
32

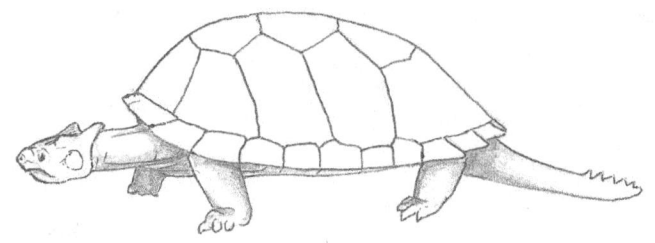

Meiolania

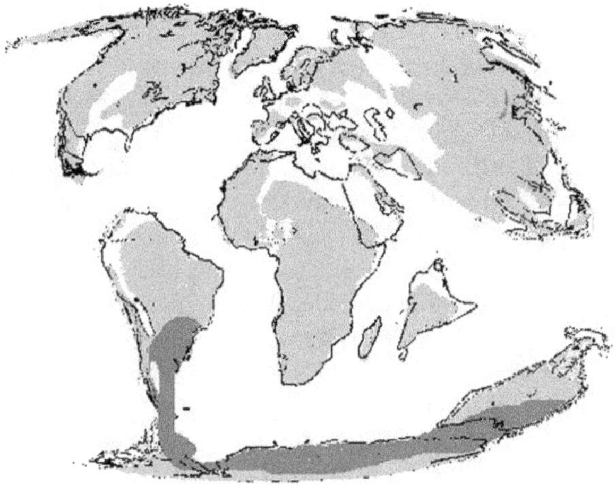

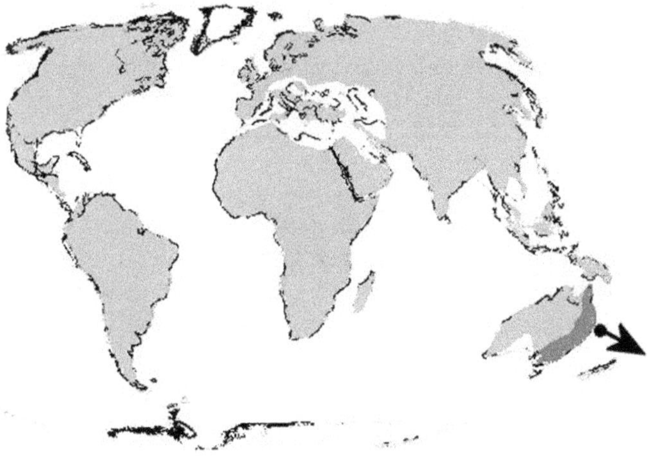

Meiolaniid evolution from 50 million years ago (top) and 100,000 years ago (bottom), showing the movement of the Lord Howe and Walpole island hotspot.

and Australia. *Niolamia argentina* of Argentina 56-34 million years ago is the origin of the family; all other meiolaniids are Australian-New Caledonian. *Niolamia* was succeeded by *Warkalania carinaminor* and *Ninjemys oweni* of Queensland 20 million and 100,000 years ago respectively. The crown group comprise *Meiolania* species from about 10 million years ago in the Northern Territories (*M. brevicollis*), and 100,000 years ago in Lord Howe island (*M. platyceps*), Queensland (unknown species) and Walpole island (*M. mackayi*). Although South America, Australia and New Caledonia were probably all exposed land masses when meiolaniids evolved, Lord Howe and Walpole islands are volcanic seamounts that date back to 5 million years ago. The hotspot that gave rise to these islands is thought to have originated off eastern Australia 80 million years ago, and has since moved eastwards as the Tasman Sea formed. It is probable that an Australian meiolaniid colonized the hotspot island in some 20-15 million years ago and as the hotspot moved eastwards, meiolaniids continued colonizing the newly formed islands, eventually colonizing the Lord Howe some 1-2 million years ago. Meiolaniids existed for at least 56 million years and despite their primitive aspect survived long after the origin of the more advanced turtles. The last of the meiolaniids was *Meiolania damelipi* from Efate Island in Vanuatu which survived until about 2,900 years ago. This remarkable late survival came to an end within 300 years of the colonisation of the island by humans, an event which appears to have led to the demise of the species.

The other members of the early fossil fauna were largely replaced by advanced forms more clearly related to the living turtles. By 100 million years ago they had all died out with the exception of species isolated on islands formed by continental drift, such as *Kallokibotion bajazidi*, the sister species to the living turtles. *Kallokibotion* was part of the Romanian Hațeg fossils identified by the eccentric palaeontologist Franz Nopcsa von Felső-Szilvás in the early 20th century. This 70 million year old fauna is remarkable for its range of species, known as the 'dwarf dinosaurs of Transylvania'. This fauna included dwarf dinosaurs that were in some cases half the size of their normal relatives, including the sauropod *Magyarosaurus dacus*, ankylosaur *Struthiosaurus transylvanicus*, hadrosaur *Telmatosaurus transylvanicus,* igaunodontids *Bihariosaurus priscus* and *Zalmoxes shqiperorum*, and the dryosaurid *Valdosaurus canaliculatus*. There were also carnivorous dinosaurs, although these are very fragmentary and not fully identified, these include the velociraptor *Elopteryx nopscsai* and the possible troodontid *Bradycneme draculae*. The carnivorous species are very poorly known, the largest species *Megalosaurus hungaricus*, is a good example – this is known just from a single tooth. Many of these carnivores appear to have been more or less normal sized. Dwarfism was not the only remarkable aspect of this fauna; it also contains the largest pterosaur ever discovered; with a 12 m wingspan the 'Hațeg Island Monster' *Hatzegopteryx thambema* must have been an amazing sight. Some of this fauna was characteristic of a scattering of islands across south and eastern Europe, for example *Bihariosaurus* was the most common dinosaur of Transylvania and is also known from France, Spain and Austria.

The tortoise fossil from Hațeg is closely tied into Nopcsa's life, which was itself almost implausible. The fossil combined his pursuit of palaeontology with his other great obsession - Albania. He was passionately interested in Albania, politically, culturally and personally; his Albanian secretary Bayazid Doda also being his lover for much of his life. The tortoise becomes central to his tragic story as he named it *Kallikobotion bajazidi* after Doda. In addition to his palaeontological pursuits, Nopsca was a backer of the guerrilla movement for Albanian independence from the Ottoman Empire, smuggling weapons into the Balkans. Although the end of the First World War saw the liberation of Albania, for Nopsca it was a tragedy as the region of Transylvania was transferred from Austria-Hungary to Romania and the Austro-Hungarian Nopsca family lost their lands. Nopsca moved to Hungary and finally to Vienna with Doda. Financial losses and the heartbreaking sale of his great fossil collection to the Natural History Museum in London were followed by descent into depression. In 1933 he drugged Doda before shooting him and then killing himself.

Kallokibotion was a small terrestrial or semi-aquatic turtle, with a carapace around 20-30 cm long. It was a primitive looking animal, lacking any significant temporal emargination. The species is represented by 31 specimens, all of which are fragmentary or isolated bones. Most of the skeleton has been found, including three skulls. Nopca thought these comprised two distinct species, *K. bajazidi* and *K. magnificum*. However, the *K. magnificum* specimen is a poorly preserved partial shell and is no longer considered to be distinct.

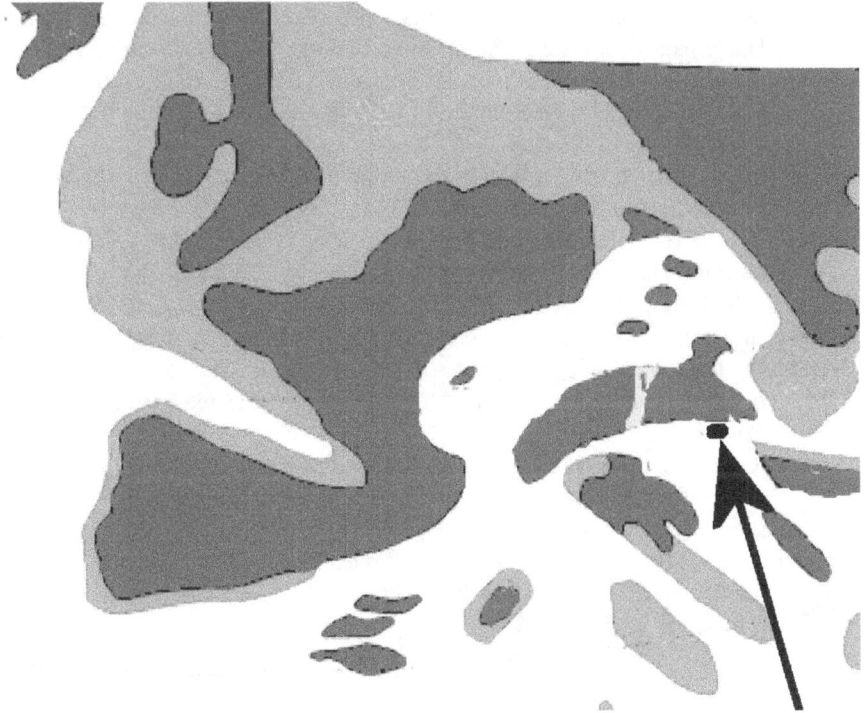

Location of Haţeg island (arrowed) within Europe 80 million years ago (dark grey – land, pale grey – shallow water).
The outline of northern Spain and France is recognisable to the left.

Hațeg island and its strange fauna were at the centre of great geological instability in Europe. 200 million years ago the continent was composed of several large northern land areas, connected by marshy low-lying areas, and southern Europe was under the sea. From 145 million years ago onwards south-western Europe was progressively pushed up and turned into dry land while the northern and eastern land areas were isolated and re-joined by fluctuating sea levels. The 80,000 square kilometres of Hațeg island was isolated from 180 million years ago until 60 and retained its early fauna for 80 million years after the primitive species such as *Kallokibotion* were replaced everywhere else by the more advanced forms, the Pleurodira, Paracryptodira and Eucryptodira.

References

Anquetin, J., P.M. Barrett, M.E.H. Jones, S. Moore-Fay & S.E. Evans. 2009. A new stem turtle from the Middle Jurassic of Scotland: new insights into the evolution and palaeoecology of basal turtles. *Proc. Biol. Soc. Lond. B* **276**: 879-886

Benton, M., Z. Csiki, D. Grigorescu, R. Redelstorff, P. Sander, K. Stein & D. Weishampel. 2010. Dinosaurs and the island rule: The dwarfed dinosaurs from Hațeg Island *Palaeogeogr. Palaeoclim. Palaeoecol.* **293**: 438-454

Csiki, Z., & M.J. Benton. 2010. An island of dwarfs - Reconstructing the Late Cretaceous Hațeg palaeoecosystem. *Palaeogeogr. Palaeoclim. Palaeoecol.* **293**: 265-270.

Csiki, Z. & D. Grigorascu. 1998. Small theropods from the Late Cretaceous of the Hateg Basin (Western Romania) - unexpected diversity at the top of the food chain. *Oryctos* **1**: 87-104

Gaffney, E.S. 1996. The postcranial morphology of *Meiolania platyceps* and a review of the Meiolaniidae. *Bull. AMNH* **229**

Gaffney, E.S., J.H. Hutchison, F.A. Jenkins Jr. & L.J. Meeker. 1987. Modern Turtle Origins: The Oldest Known Cryptodire. *Science* **237**: 289-291

Gaffney, E.S. & F.A. Jenkins. 2010. The cranial morphology of *Kayentachelys*, an early Jurassic cryptodire, and the early history of turtles. *Acta Zool.* **91**: 335-368

Grigorescu, D. 2005. Rediscovery of a 'forgotten land': The last three decades of research on the dinosaur-bearing deposits from the Hațeg Basin. *Acta Palaeont. Rom.* **5**: 191-204

Joyce, W.G., S.G. Lucas, T.M. Scheyer, A.B. Heckert & A.P. Hunt. 2009. A thin-shelled reptile from the Late Triassic of North America and the origin of the turtle shell. *Proc. R. Soc. B* **276**: 507–513

Rougier, G.W., M.S. de la Fuente & A.B. Arcucci 1995. Late Triassic Turtles from South America *Science* **268**: 855-858

Sterli, J., M.S. de la Fuente & G.W. Rougier. 2007. Anatomy and relationships of *Palaeochersis talampayensis,* a Late Triassic turtle from Argentina. *Palaeontographica* A **281**: 1-61

Sterli, J. & M.S. de la Fuente. 2011. A new turtle from the La Colonia Formation (Campanian–Maastrichtian), Patagonia, Argentina. *Palaeontology* **54**: 63-78

Sterli, J., J. Muller, J. Anquetin & A. Hilger. 2010. The parabasisphenoid complex in Mesozoic turtles and the evolution of the testudinate basicranium. *Can. Journ.*

Earth Sci. **47**: 1337-1346

White, A.W., T.H. Worthy, S. Hawkins, S. Bedford & M. Spriggs. 2010. Megafaunal meiolaniid horned turtles survived until early human settlement in Vanuatu, Southwest Pacific. *PNAS*

5. Evolution of the pleurodires

Present-day pleurodires are restricted to the Pelomedusidae of Africa, the Podocnemidae of South America (with an isolated species on Madagascar), and Chelydridae of Australia and South America. Early in their evolutionary history though Pleurodires appear to have been considerably more diverse, especially in coastal areas. Several extinct families are known, including Bothremydidae, Araripemydidae and Euraxemydidae.

The pleurodire neck retraction was possible 200 million years ago in *Platychelys* of Europe and *Notoemys* and *Caribemys* of southern Gondwana. These had neck vertebrae similar in arrangement to those of the living Chelidae: vertebrae 2-4 are 'opisthocoelous' (concave posteriorly), vertebra 5 biconvex (concave at both ends), vertebra 6 'procoelous' (concave anteriorly), 7 biconcave and 8 biconvex. This gives a high degree of neck

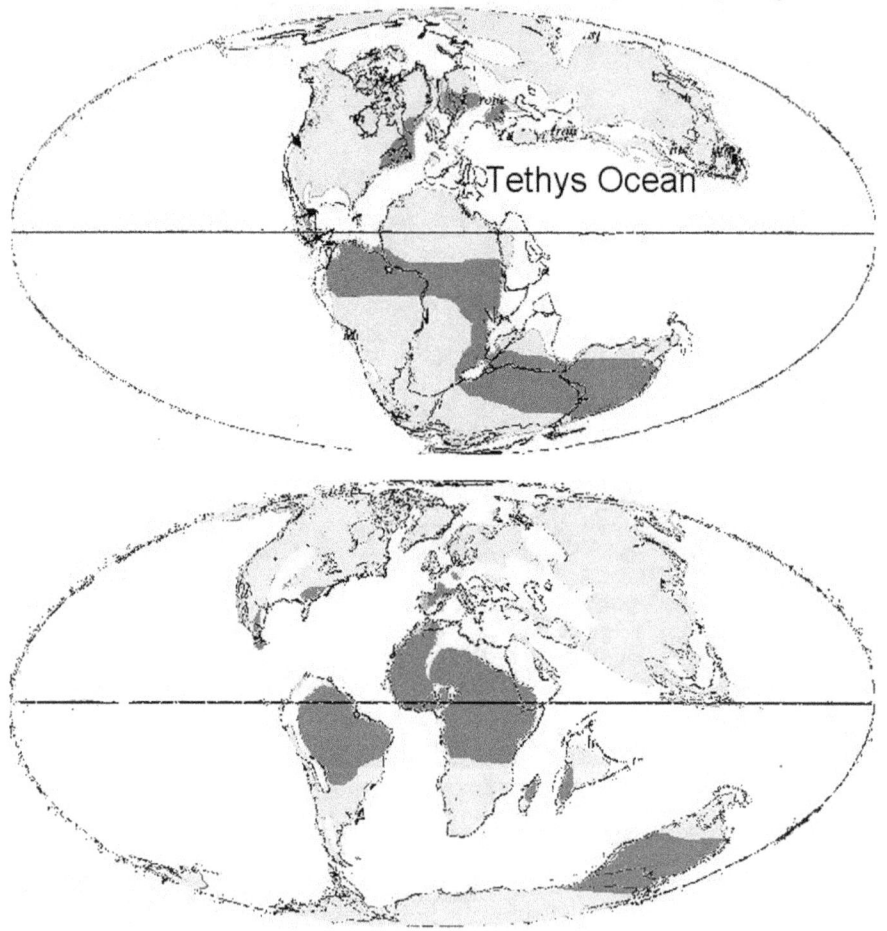

Distribution of pleurodires 150 and 66 million years ago.

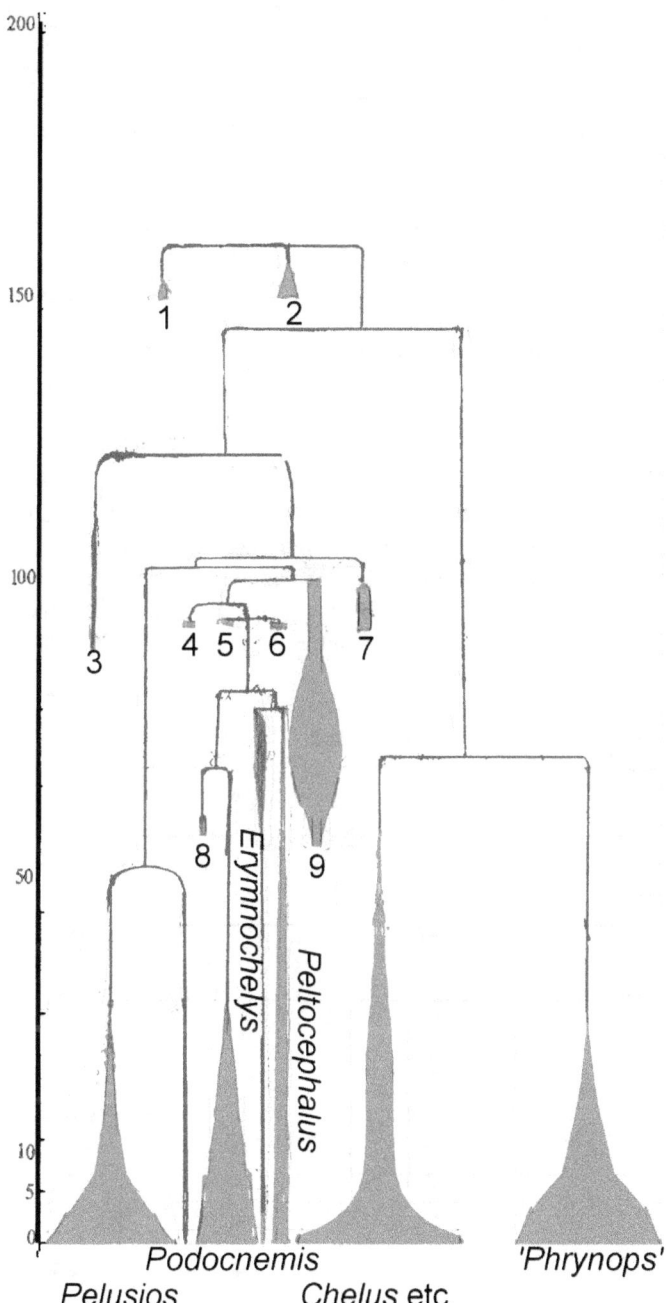

Pleurodire phylogeny
1 - Dortokidae; 2 - Platychelydae; 3 - Araripemydidae; 4 - *Brasilemys*, *Hamadachelys* and *Portezueloemys*; 5 - Euraxemydidae; 6 - early podocnemidinae; 7 - Euraxemydidae; 8 - *Cerrejonemys*; 9 - Bothremydidae .

mobility at the base of the neck and around the 5th vertebra. The neck configuration was later modified in the pelomedusoides moving mobility closer to the head (vertebra 2 biconvex, all others procoelous). Advanced pelurodires with linked processes on the neck vertebrae date from 145 million years ago with the Dortokidae and the Eupleurodires. The earliest pleurodires, the 145 million years old *Notoemys zapatocaensis*, *N. oxfordiensis* (also known as *Caribemys oxfordiensis*) and *N. laticentralis*, were highly hydrodynamic, flattened freshwater turtles. These were found around the margins of the west of the Tethys Ocean, with fossils ranging from Argentina to Cuba and Colombia from 156 to 135 million years ago. These were closely related to the European *Platychelys* from the eastern end of the Tethys, which probably lived in coastal lagoons. Thus around this time early pleurodires were found all around the margins of the Tethys, probably living in coastal waters and lagoons.

Of the more advanced pleurodires the Dortokidae were exclusively European. These coastal pleurodires appear first in the west of Europe, in Spain 130 million years ago. They seem to have spread eastwards gradually, being found in France and Austria 84 million years ago and Romania 70 million years ago. The islands that made up Europe at this time were in a shallow sea well suited to aquatic turtles. By 70 million years ago the islands were joining as the low-lying areas of Europe became exposed by falling sea-levels. This seems to have precipitated the decline of the dortokids; they were last recorded in Spain from that time. The best known genus, *Dortoka* did not survive after 65 million years ago. By 55 million years ago the Dortokidae were replaced throughout Europe by the living families Podocnemididae and Emydidae. The last dortokid known was the Romanian *Ronella botanica*.

The Eupleurodires probably evolved from the early *Notemys* pleurodires from the west of the Tethys. While those in what was to become Europe gradually declined and were displaced by the Emydidae, those of Gondwana thrived. These divided into two

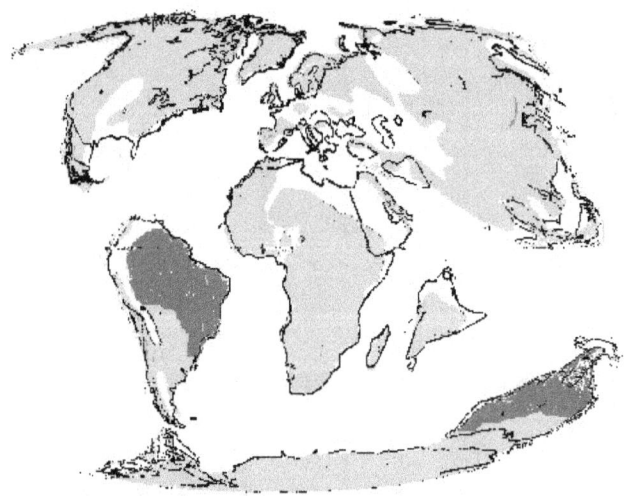

Distribution of Chelidae from 70 million years ago.

distinct lineages – the Chelidae and the Podocnemidodea. Chelidae are the Australian and South American side necks, whilst the Podocnemidodea include the Podocnemididae, Pelomedusidae and a diverse range of ancestral fossil groups. Within the Chelidae there are two clearly different groups: the long-necked and the short-necked species. These are found in both Australasia and South America and were originally thought to indicate that chelids specialised into these two forms at a time when these continents were still in contact. However, recent DNA studies show that these groups are not valid and that they are the result of convergences between the Australasian and South American faunas. The new interpretation is that the chelids were divided into two major geographical groups when the continents divided between 65 and 50 million years ago, separating South America from Australia-Antarctica. For the past 50 million years ago South American and Australian turtles have been evolving in isolation, in both continents some turtles have evolved long necks and other short necks. Neck length is associated with their feeding strategies, with long-necked species evolving as specialised fish predators.

Long-necked turtles all have remarkably long necks and relatively small, elongate, streamlined skulls. They use a gulping method of prey capture, which requires the head to be small for fast protraction and the body large to prevent it being pushed backwards during the predatory strike. For the neck to be shot forwards in the strike the neck movement muscles (*longissimus dorsi*) along the vertebrae under the shell have to be exceptionally large, which is associated with expansion of the neural bones of the shell and enlargement of the plastral buttresses. These fish predators all have flat heads, with good binocular vision.

In South America long-necked forms evolved independently in the subfamilies Chelidinae and Hydromedusinae (represented today by *Chelus* and *Hydromedusa* respectively). Hydromedusinae also include the fossil genus *Yaminuechelys*. This is known from around 60 million years ago in Patagonia (Argentina), mainly around the Rio Negro. *Yaminuechelys gasparinii* was a long-necked chelid with well-developed lateral cheek emargination, which would have been associated with powerful jaw and neck musculature typical of an ambush predator of fish. It seems to have been closely related to *Hydromedusa* in that it has widened internal nostrils, caused by reduced development of the palatine bones. It is more primitive than *Hydromedusa* as its skull has less lateral emargination, larger hyoid bones in the throat, presence of lateral mesoplastral bones in the plastron and a more primitive arrangement of sacral and neck vertebrae.

Hydromedusa

Hydromedusa is known from 56 million years ago in Argentina. The are two living species, the Brazilian snake-necked turtle *H. maximilliani* of south-eastern Brazil and the South American snake-necked turtle *H. tectifera*, which lives in the same area but extends further south than the Brazilian, into Argentina. The Brazilian snake-necked contains two genetic forms (as indicated by mitochondrial DNA), one in the east (Rio de Janeiro – Minas Gerais) and one in the west (Sao Paulo). These groups seem to have diverged somewhere between 16 and 8 million years ago. This corresponded to the geological upheaval forming the Serra do Mar and Serra da Mantiqueira mountains that divide the range of the species. Periods of dry conditions from around 30 million years ago to 5 million years ago caused wetland-associated species to be restricted to valleys. The snake-necked turtles are found in shallow water courses and their range would have been fragmented at this time, with geological changes combining with the climate to separate east and west populations.

Chelidinae

Chelidinae include the living genera *Chelus* (the matamata turtles), *Acanthochelys* (South American side-necked swamp turtles), *Mesoclemmys* (Gibba turtle), the '*Phrynops*' group (toad-headed turtles), *Platemys* (twisted-necked turtles), *Rhinemys* (red-headed sideneck turtle) and the fossils *Bonapartemys*, *Lomalatachelys*, *Prochelidella*, *Palaeophrynops*, *Parahydraspis* and *Linderochelys*. Chelid fossils appear 110 million years ago in Patagonia, and *Lomalatachelys* is one of the earliest genera (90 million years ago).

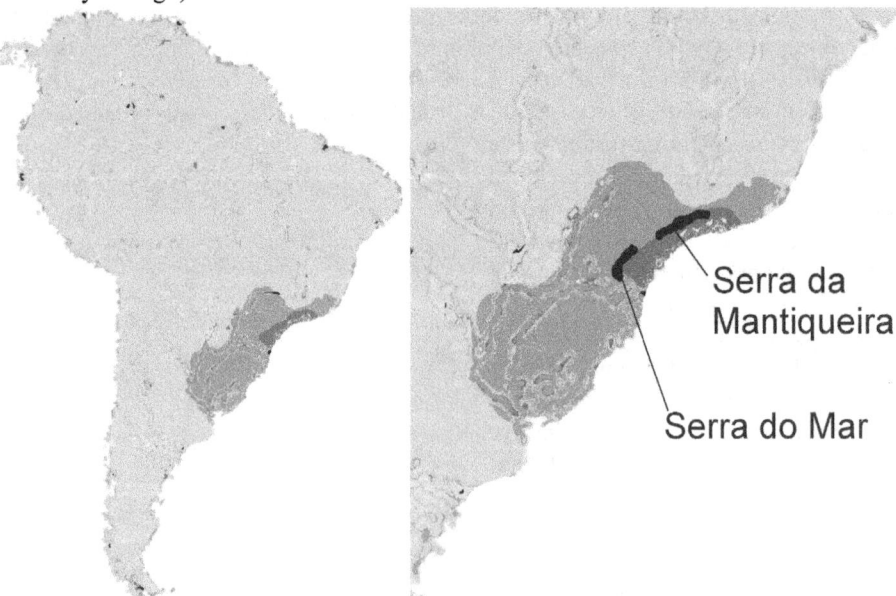

South America showing the small range of *Hydromedusa maximilliani* bordered by the more widespread *H. tectifera*.

Chelus contains the most bizarre of all turtles, the matamata *Chelus fimbriatus.* The matamata lives throughout the Amazon Basin and is found in most of the major rivers of north-eastern South America. In these river courses it lives in slow moving, muddy waters where it lurks, waiting for prey. It feeds on a variety of small aquatic animals which it catches by a strong sucking action. The strange skull shape is adapted to allow the floor of the mouth to be depressed to an unusual extent; this creates very powerful suction. The matamata is one of the most specialised ambush predators, and can remain immobile underwater for hours by absorbing oxygen through the skin of the throat and cloaca. All aspects of their biology are tied to their life in swamps; the y have a particularly variable incubation duration with a normal duration of 80 days, but an extreme of over 200 due to delayed development of some clutches as an adaptation to fluctuating water levels. Although it is an extremely specialised form *Chelus* appears to have evolved at least 15 million years ago. Fossils are known from Venezuela and western Brazil about 15-10 million years ago (*C. lewisi*) and Colombia about 15 million years ago (*C. colombiana*). The ancestor of these species was probably widespread throughout the main rivers of South America. About 15-11 million years ago geological upheaval raised the Eastern Cordillera mountains of Colombia which would have divided the South American rivers into western (Magdalena, Maracaibo and Atrato) and eastern (Orinoco and Amazonas) groups. This would have isolated *C. colombiana* in the Magdalena Valley, and left *C. lewisi* and *C. fimbriata* in the Orinoco-Amazon basin.

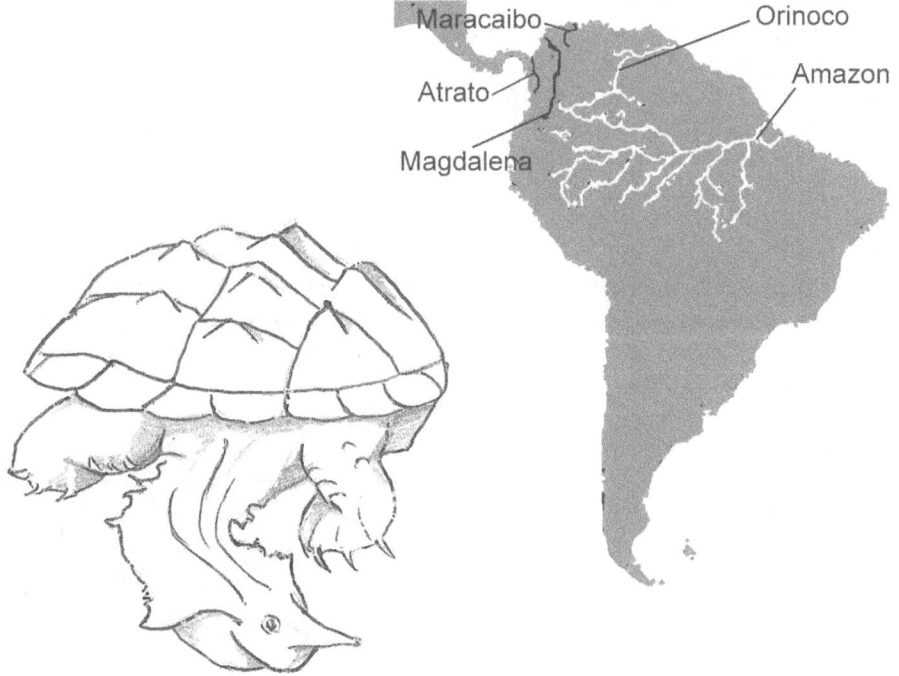

Chelus and map of South America showing the isolation between the western Maricaibo-Atrato-Magdalena and eastern Orinoco-Amazon river drainages

The twist-necked turtle *Platemys platycephala* is found in the Orinoco and Amazon from Brazil to Peru. This is similar to the *Acanthochelys* swamp turtles. The eastern population is regarded as distinct (*platycephala*), separated from the black-backed twist-neck *melanota* from Ecuador and western Peru. These turtles are found in shallow pools or small lakes in forest habitat. In the dry season these water bodies dry out and the twist-necked turtle buries itself, waiting for rain. Eggs are laid in this season and in order to survive the dry conditions these are unusually large (being up to 30% the length of the adult turtle). This minimises water loss, allowing them to survive 4-5 months of dry conditions.

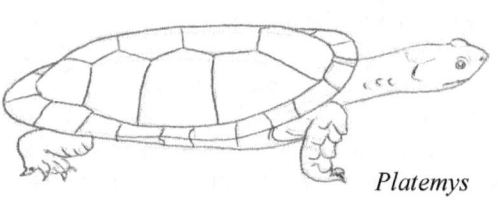

Platemys

The similar *Acanthochelys* genus includes four species. The Pantanal swamp turtle *Acanthochelys macrocephala* is found in the upper Rio Mamoré of Brazil and Paraguay. Chaco side-necked turtle *A. pallidipetoris* occurs in the Chao of Argentina, Paraguay and Bolivia. The Brazilian radiolated swamp turtle *A. radiolata* lives in the coastal rivers of Brazil. The spiny-necked swamp turtle *A. spixii* is found further south in Brazil and in Uruguay, mainly in coastal lagoons. *Acanthochelys* are less specialised than *Platemys* and are predatory animals of marshes, feeding largely on snails. Fossils similar to *Acanthochelys* are known from Patagonia 100 million years ago.

Acanthochelys

Many South American side-necks were described in the toad-headed sideneck turtle genus *Phrynops*; this is now split into six genera: *Batrachemys*, *Bufocephala*, *Mesoclemmys*, *Phrynops*, *Ranacephala* and *Rhinemys*. Fossils of this group are known from Patagonia 100 million years ago, but are not definitely identifiable as members of *Phrynops* until 23 million years ago when fossils from Sao Paulo, Brazil can be attributed to the superspecies *Phrynops* 'geoffroanus'. This genus includes at least five recent species and the Brazilian

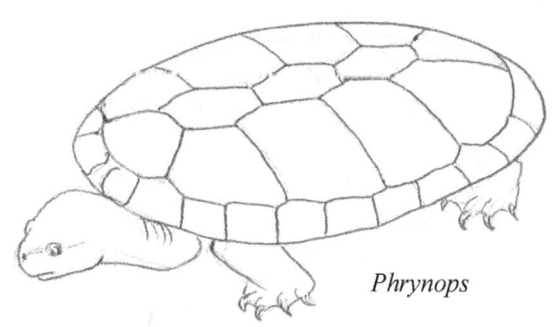

Phrynops

44

fossil *Parahydraspis paranaensis*. True *Phrynops* are restricted to Geoffroy's side-neck *P. geoffroanus,* Peter's *P. tuberosus,* the spotted-bellied side-neck *P. hilarii* and William's South American side-neck *P. williamsi*. *Phrynops* is sister group to the other genera: *Rhinemys, Mesoclemmys, Ranacephala, Bufocephala* and *Batrachemys*. *Phrynops* favour slow moving waters and are mainly carnivorous.

Several species are related to the toad-heads. The first of these are the genera *Rhinemys* and *Mesoclemmys* which seem to originate from the middle-reaches of the Amazon and Orinoco rivers, then spread widely. *Rhinemys* contains only the red-headed sideneck *R. rufipes* of the mid to upper courses of the Amazon. It lives in a range of water types, fast or slow flowing, clear or turbid. It is omnivorous, and seems to be especially attracted to palm fruits. *Mesoclemmys* comprises the Gibba turtle *M. gibba* which is widespread in the Orinoco and Amazon and on the island of Trinidad, and the toad-heads which are sometimes referred to as *Batrachemys*. The Gibba turtle lives mainly in forest pools and is largely carnivorous. The *Batrachemys* toad-heads are more diverse, containing the common toad-head of Surinam, French Guiana and the northern end of Brazil *B. nasuta;* Western Amazon toad-head *B. heliostemma* from the upper tributaries of the Amazon, Amazon toad-head *B. raniceps* of the Amazon; Dahl's toad-head *B. dahli* of a small part of Colombia (the Rio Sinu valley); and the sibling species Zulia toad-head *B. zuliae* of north-western Venezuela*;* and the tuberculate toad-head *B. tuberculata* of eastern Brazil. Dahl's is the only side-neck found east of the Andes. Zulia's and the tuberculate toad-head may be related although the ranges of these species are widely separated. Zulia's is very restricted, and largely isolated from the range of other species by an arid region that was much wetter in the past. These species are all predators of slow moving, muddy waters. The most recently discovered member of this group is the Piaui side-neck *M. perplexa*. This was described in 2005 from animals from north-east Brazil (Ceara and Piaui states).

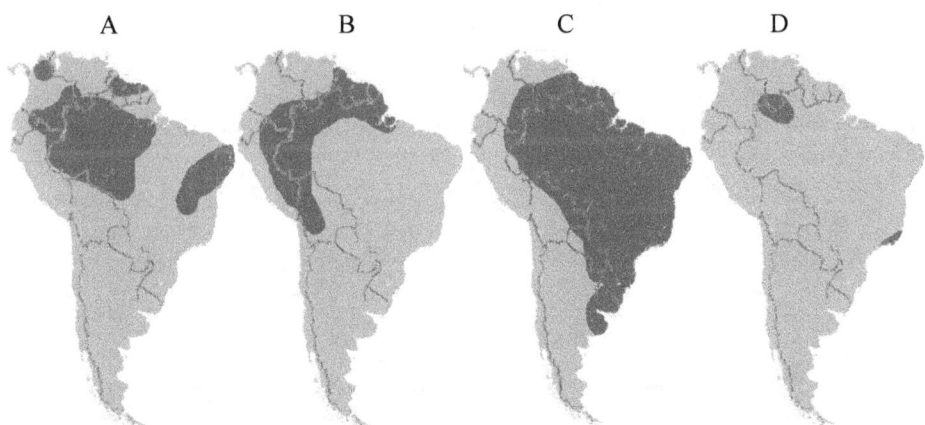

Range of toad-heads:
A) *Batrachemys*, B) *Mesoclemmys* (north) and *Bufocephala* (south), C), *Phrynops,*
D) *Rhinemys* (north) and *Ranacephala* (south).

More derived relatives spread from an Amazonian origin into the Brazilian coastal rivers (*Ranacephala*) and the southern parts of the Amazon region (*Bufocephala*). *Ranacephala* is Hoge's side-neck *R. hogei* from the coastal rivers of south-east Brazil. This species is only partially aquatic, living in and near small lakes. *Bufocephala* is Vanderhaege's toad-head *B. vanderhaegei* from southern Brazil to northern Argentina. This species lives in shallow lakes and is mainly carnivorous.

Chelodininae

The Chelodininae are the chelids of Australia and New Guinea (both short- and long-necked). All molecular studies agree that this is a real group, in contrast to many morphological studies that suggest that the long-necked Australian and South American species group together. Within the Australasian species three distinct groups can be identified: the long-necked species (*Chelodina* and related genera), the *Emydura* group, and *Pseudemydura*. This latter genus appears to be the most primitive of the living short-necks and is sometimes put in its own subfamily (Pseudemydurinae). The *Emydura* group comprises the short-necked Australian snapping turtles *Emydura* and *Myuchelys*, *Rheodytes*, *Elusor* and *Elseya*. The relationships of these, and even the number of species in them, is still not clear. *Myuchelys* may be the first of these to have separated from the main groups, followed by *Rheodytes* and finally *Elseya* and *Emydura*.

Long-necked Australian chelids comprise *Chelodina*, *Macrochelodina* and *Macrodiremys*. These are regarded as genera or subgenera. Fossils of these types of turtles (not easily identifiable to genus or subgenus) are known from 20 million years ago to the present. These long-necked species all have similar adaptations of the skeleton and muscles. The most extreme are the *Marochelodina* species in which the extremely long neck has expanded retrahens capitus collique and longissimus dorsi muscles. The powerful muscles require the shell to be strengthened to withstand the forces created in rapid neck movement, this strengthening is apparent in the enlarged struts to the bridges of the narrow plastron. In association with this powerful musculature the neck vertebrae are elongate and robust, and the atlas/axis complex which connects the skull to the back-bone is completely fused. The jugal bone of the skull is wide, and the squamosal is large to support the digastricus maxillae muscle.

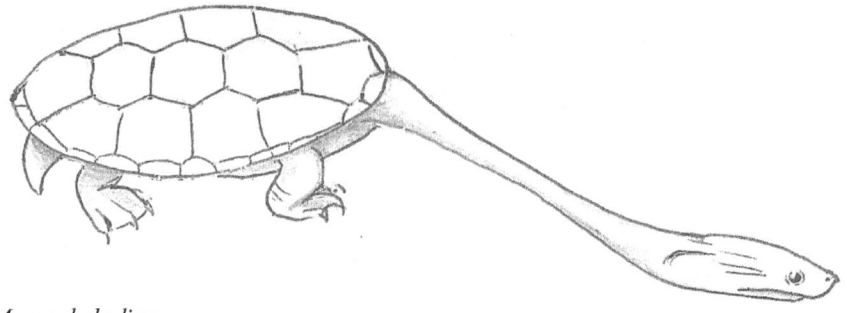

Macrochelodina

In *Chelodina* there is a small anterior strut to the plastral bridge, although in *C. novaeguineae* this is enlarged towards the condition seen in *Macrochelodina*. The neck musculature of *Chelodina* is also not as developed as in *Macrochelodina*. In association with this the neural bones of the shell never form a complete series; it has been suggested that the tendency to lose neural bones in many pleurodires may be the result of a change in development rates of different parts of the body. Under this suggestion the neural bones retain their juvenile form, whereas the costals develop normally, eventually excluding the neurals altogether. The extreme of flat shell, small buttresses and wide plastra are seen in *C. steindachneri* and *C. longicollis*. Similarly, in the skull the jugal and squamosal are small.

In *Macrochelodina* the extreme development of the musculature is seen in the sandstone longnecked turtle *M. burrungandjii* where the exceptionally well developed neck musculature is associated with well developed neural bones, which are significantly reduced in most other chelids. These 'oblong snake-neck turtles' are found in the east and north of Australia and on New Guinea, comprising five living species and at least two fossils (*M. alanrixi* from around 40 million years ago and *M. insculpta* from within the last 5 million years, both from Queensland). This last species is probably related to the giant snake-necked turtle *M. expansa*. The giant snake-neck of south-east Australia is the largest of the '*Chelodina*' species, reaching up to 48 cm. This is the only species in this group to live in the south-eastern region; it was probably more widespread in eastern Australia in the past but during the last Ice Age cold, dry conditions would have contracted the range, leaving this species isolated from the rest of the *Macrochelodina* species in the north. There is also a population on Fraser Island off the east coast of Australia which may have been introduced by humans at an unknown date. The northern species are Parker's longneck *M. parkeri* (New Guinea), northern longneck *M. rugosa* (Northern Australia, New Guinea) and the closely related sandstone longneck *M. burrungandjii* (Northern Australia) and Kimberley longneck *M. walloyarrina* (Nortwestern Australia). There is also Kuchling's longneck *M. kuchlingi* of north-western Australia but this is based on a single specimen and is of uncertain validity. The northern longneck has two subspecies: *M. r. rugosa* in Australia and *M. r. siebenrocki* in New Guinea. This range surrounds the Gulf of Carpentaria which would have been a lake surrounded by dry land 55,000–12,000 years when low sea levels exposed what is now the sea-bed of the Torres Strait. At this time northern Australia and New Guinea would have been joined, allowing colonisation of New Guinea by the Australian sidenecks.

The northern snakeneck is unusual in that females lay their eggs in the bottom of temporary muddy pools. These eggs do not start to develop until the dry season when the pools dry out. The hatchlings remain in the nest in the now baked dry mud until the next rains soften the soil. Sometimes this may be 2 years after the eggs were laid. The period of resting before egg development starts (diapause) is needed for development of most sideneck species. In the case of the giant snake-neck this may extend total incubation to 664 days, although the normal limit is probably 192 days.

The northern and sandstone longnecks will hybridize to produce fertile offspring wherever the ranges of the two species overlap. Mitochondrial DNA appears to show that genes from the sandstone species have spread through the range of the northern species in Arnhem Land, resulting in some morphological differences which have been considered to indicate distinct subspecies: *M. b. burrungandjii* and *M. b. walloyarrina*.

Chelodina are found from all of Australia except the south-west. Steindachner's turtle *C. steindachneri* lives in throughout Western Australia. This is a strange habitat for an aquatic turtle. It survives by aestivating in dried out water bodies. It is able to store considerable quantities of water in accessory bladders and so can survive for up to two years until rains return. The first species is the only one from the south: the common snake-neck *C. longicollis* (Southeast Australia). This species has been present in this area for the past couple of million years as indicated by fossils from the Naracoorte Caves of South Australia. It is an adaptable species and there have been accidental introductions into Tasmania. The northern species form a group, most of which are restricted to New Guinea but which probably originated in north-western Australia. These northern species comprise: Steindachner's snake-necked *C. steindachneri* (Western Australia), Pritchard's snake-neck *C. pritchardi* (New Guinea), Roti Island snake-neck *C. mccordi* (Roti Island, Indonesia), Reimann's snake-neck *C. reimanni* (New Guinea) and the closely related species pair of New Guinea snake-neck *C. novaeguineae* and Cann's snake-neck *C. canni* (Northern Australia). These last two species are very similar and probably had a recent origin from an ancestral species distributed throughout the Gulf of Carpentaria. This would have diverged into two isolated forms some 15,000 years ago when this area was dry land (as with the northern longneck turtle subspecies). New Guinea snake-neck fossils have been reported from 5 million years ago from Queensland; these may represent the ancestral widespread form. The Roti Island snake-neck has three subspecies: *C. r. mccordi* from the west of the island, *C. r. roteensis* from the east,

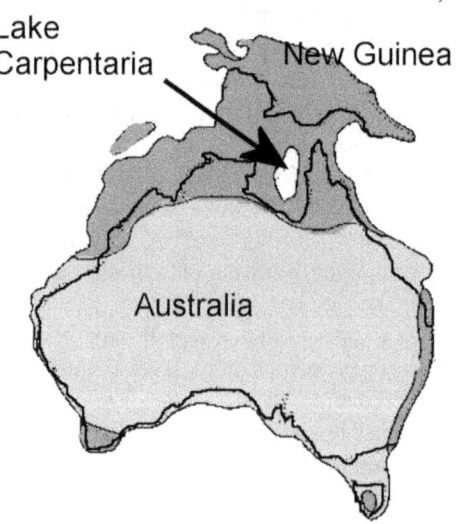

Australia and New Guinea 50,000 years ago showing Lake Carpentaria in the dry Torres Strait. Light grey – semi-arid land, dark grey – wetter habitats.

and *timorlestensis* from Timor-Leste. There is also Gunalen's snake-neck *Chelodina gunaleni* which was described from a very small area of south-western New Guinea but may be a local form of the New Guinea snake-neck. The relationships between these various species are somewhat unclear as at least some hybridize in the wild; the common snake-neck and Cann's snake-neck frequently hybridize in the Styx River in Queensland and these hybrids are fertile. Similarly, Cann's and the New Guinea snake-neck are visually recognisable but cannot easily be told apart by molecular methods.

The narrow-breasted snake-necked turtle *Macrodiremys colliei* (or *oblonga*) from the south-west of Australia (especially around the Swan river) has much larger *longissimus dorsi* muscles than in any other Australasia chelid. This has a more or less complete series of neurals. It is probably related to the *Chelodina* species.

Pseudemydura has a single living species, the Western swamp turtle *P. umbrina* which is found only in two sites near Perth in Western Australia. This species was known from a single specimen collected in 1839 until 1953 when it was rediscovered in the wild. It is found in seasonal water-bodies and spends much of its time buried on land, aestivating during long dry seasons. The eggs pass a similarly long incubation, hatching after several months as rains start to restore marshy conditions. These adaptations to harsh conditions mean that the distribution is naturally limited and populations small. This most isolated of all Australian turtles (which are almost all northern or eastern) is probably a relict species which was formerly

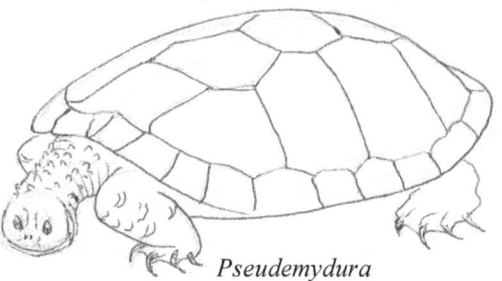
Pseudemydura

more widespread and abundant when climatic conditions were more favourable. It seems to have been restricted to this area of Western Australia for a considerable period of time; fossils similar to this species are known from about 15 million years ago. Populations have been further reduced in recent decades by predation by introduced foxes and habitat damage caused by rabbits. In 1962 the only sites for the species were protected by the creation of the Ellen Brook and Twin Swamps Nature Reserves. These cover 220 hectares and in the 1960s the total population was estimated at 30 individuals in Ellen Brook and 200 in Twin Swamps. By the 1980s the latter had declined to only four animals; in 1987 the total number was estimated at fewer than 50 individuals. Captive breeding has allowed reintroductions to take place. This and maintenance of habitats have resulted in a wild population of about 250 (with a further 200 in captivity).

The next group of short necks to evolve was *Myuchelys*. These are the Australasian helmeted turtles or saw-shelled turtles. They mainly live in small rivers (tributaries and headwaters) in northern and eastern Australia and New Guinea. Five species are known at present: the saw-shelled turtle *M. latisternum* (northern and eastern Australia), Bellinger River snapper *M. georgesi* (Bellinger River, eastern Australia), Manning River snapper *M. purvisi* (Manning river, southeastern Australia), Namoi River snapper *M. bellii* (Gwydir, Namoi, MacDonald rivers and Bald Rock Creek in southeastern Australia) and the New Guinea snapper *M. novaeguineae* (north and south New Guinea). The New Guinea snapper resembles *Elseya* and was included in this genus until recently; DNA analysis clearly places it in *Myuchelys*. *M. schultzei* may be a subspecies of the New Guinea snapper. It is probable that the ancestral *Myuchelys* species resembled the south-eastern Belling and Manning River species, and lived in similar habitat to these species (clear and fast flowing rivers with stony beds). These very restricted species represent remnants of the ancestral species. This was followed by more adaptable forms that moved northwards, giving rise to the saw-shell throughout the coastal east and north, the Namoi River snapper of the east and ultimately the New Guinea snapper.

More derived than these are the geographically restricted genera of eastern Australia *Elusor, Rehodytes* and *Birlimarr. Elusor* contains only the Mary River turtle *E. macrurus* of the Mary River of Queensland. This species has a moderately long neck and a distinctively large tail. This is unusually long and is laterally compressed, giving it a rudder-like function. *Rheodytes* contains only the Fitzroy River turtle *R. leukops* of the Fitzroy River, Mackenzie and Dawson rivers of Queensland. This has massive jaws which would be well adapted for crushing shells, but its diet seems to be largely opportunistic, including insect larvae and massive quantities of vegetation. It is possible that its highly restricted range (less than 10,000 km²) has an ecology that is rather different from the environment in which it evolved, and that naturally it should be a snail-feeding

Rheodytes

specialist. It may be related to *Emydura*. As with many Australian turtles this species can obtain oxygen from the water by absorbing it through the cloaca, giving it the inelegant local name of the "bum-breathing turtle". This characteristic is also found in the Mary River turtle. There is a fossil species from much the same area, *R. devisi*, from sometime within the past 5 million years. *Birlimarr* is even more restricted, being known only from a single fossil species *B. gaffneyi* from 12 million years ago.

The most derived members of this group are *Emydura* and *Elseya*. *Elseya* are the 'Australasian snapping turtles' which are restricted to the north and east of Australia and New Guinea and evolved around 10 million years ago. There is a split in the genus: northern snapping turtle *E. dentata* (northern Australia) and Alligator River snapper *E. jukesi* (South Alligator river, north Australia)

Emydura

50

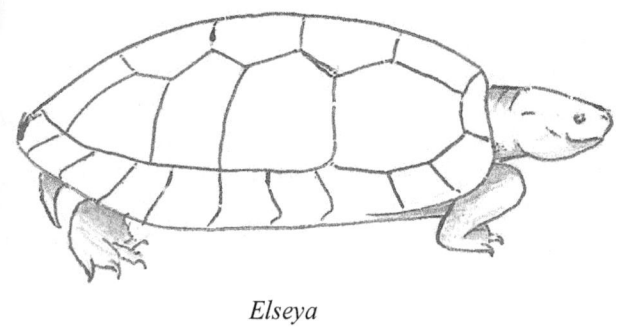

Elseya

are distinct from the rest of the genus. The main part of the genus comprises Lavarack's snapper *E. lavarackorum* (northern Australia) and the pink-bellied snapper turtle *E. branderhorsti* (south New Guinea) and a grouping of the southern snapper *E. albagula* with Irwin's turtle *E. irwini* (Burdekin River, Queensland) and Stirling's turtle *E. stirlingi* (Johnstone river, Queensland). Lavarack's snapper was known only as a fossil until 1997, when an un-named form of *Elseya* from the Nicholson river drainage was realised to be identical to the fossils. These fossils date from around 5 million years ago from Queensland. One species is known only from these fossil deposits: *E. nadibajagu*, which seems to be related to Itwin's turtle. Lavarack's snapper is found in the rivers that flow into the Gulf of Carpentaria. This seems to be very similar to the pink-bellied snapper from southern New Guinea. Several '*Chelymys*' species previously described from fossils are now placed in *Elseya*: *Chelymys uberima, C. arata, C. antiqua* and *Pelecomastes ampla* are all *E. uberima*, from Queensland within the past 2.5 million years. This species is related to Lavarack's snapper. It is probable that this genus has always been restricted to north-east Australia and that early in the evolution of the genus three geographical forms evolved: the north-western (northern and Alligator River snappers), north-eastern (Lavarack's and pink-bellied snappers) and the eastern (Irwin's, Stirling's and southern). The north-western group is distributed around the Gulf of Carpentaria and as with other Australian chelodines must have diverged into separate Australian and New Guinea species in the past 12,000 years as these areas became isolated by rising sea levels. The northern snapper is the least carnivorous of the Australian side-necks, the adults being almost completely herbivorous.

The origin and distribution of many of the eastern Australian turtles was strongly influenced by Ice Ages 30,000 to 10,000 years ago. From about 30,000 years ago cooling climates led to the build up of ice in Australia, reaching a maximum extent 18,000 years ago. Unlike the Ice Age climate of the northern hemisphere the ice sheets were not extensive in Australia, covering only about 25km^2, but a much wider area would have been subject to winter frost and cold, dry conditions. 12,000 years ago this aridity reached its maximum over south coastal and central Australia. At this time 81% of lakes were at extremely low levels or had dried out completely, resulting in the extinction of most of the freshwater fauna. The Murray-Darling catchment, now home to *Emydura* species was almost completely dry 18-12,000 years ago. The few populations that survived were isolated and diverged into distinct forms (species and subspecies). The retreat of the glaciers 10,000 years ago allowed recolonisation by aquatic animals, some spreading from the less arid north, others expanding from their isolated remnant populations.

Emydura are the 'Australasian river turtles'. They are known from 15 million years ago to the present. Fossils are known from Tasmania but there are no native chelonians there now as a result of extensive ice cover during the last Ice Age. About 18,000 years ago Tasmania would have been covered by some 700m of ice. Since the melting of the ice cap Tasmania has been isolated from the mainland by 240 km of ocean, preventing turtles recolonising the area. There are five living species from Australia and New Guinea: painted sideneck turtle *E. subglobosa* (New Guinea), Australian big-headed turtle *E. australis* (north-west Australia), Murray River turtle *E. macquarii* (east and south-eastern Australia), Tanybaraga turtle *E. tanybaraga* (northern Australia) and Victoria short-necked turtle *E. victoriae* (north-western Australia). Of these the Victoria species is a sister species to the Tanybarga turtle, and these are related to the big-headed turtle and to the painted sideneck. The most isolated species is the Murray River turtle. This contains four subspecies: Murray River turtle *E. m. macquarii*, Krefft's turtle *E. m . krefftii*, Fraser Island short-neck turtle *E. m. nigra* and the Cooper Creek turtle *E. m. emmotti*. Of these the Fraser Island and Krefft's form are sister species, and these are related to the Murray River and then to Cooper Creek. The painted sideneck contains two subspecies: red-bellied short-necked turtle *E. s. subglobosa* (New Guinea) and Worrell's short-necked turtle *E. s. worrelli* (northern Australia). Early in the evolution of the genus an ancestral river turtle divided into north-western and south-eastern populations. The north-west form diverged over the last 6,000 years when New Guinea was isolated from Australia (giving rise to the painted sideneck), followed by more recent isolation in of the north-western (big-headed) and mainly northern (Tanybaraga) species as increasing aridity isolated the different river systems. The south-east form gave rise to the Murray River turtle which is known from the east of Australia from 5 million years ago to the present.

It seems that the strike and gape mode of feeding has evolved twice in the Australasian chelids; in *Macrochelodina* using the muscles that run between the neck and the head (*retrahens capitus collique*), giving a quick head action, and separately in the narrow-breasted snake-necked turtle *Macrodiremys* which uses the muscles at the base of the neck (*longissimus dorsi*) muscles, striking at high speed with a straightened neck. Other *Chelodina* species lack the musculature for such rapid movements and are suck and gape feeders instead, swimming after their prey rather than being ambush predators.

The narrow-breasted snake-necked turtle of the Swan Coastal Plain of Western Australia is one of very few species to have been studied for vocal communication. 17 distinct vocalizations have been identified, comprising clacks, clicks, squawks, hoots, short chirps, high short chirps, medium chirps, long chirps, high calls, cries or wails, hooos, grunts, growls, blow bursts, staccatos, a wild howl, drum rolling and a sustained mating call. It is thought that an unusual vocal repertoire has evolved in this species as it normally lives in muddy or tannin-stained waters where visibility is very poor.

Podocnemidodea

The podocnemidods first appear as the freshwater Araripemydidae from

Gondwana. The earliest of these were the South American *Araripemys barretoi* and the African *Teneremys lapparenti* from Niger, some 112 million years ago. At this time South America retained a narrow connection with Africa, between what is now Brazil and Niger. Thus the earliest podocnemidods come from a small geographical area. This family spread northwards, as shown by the 65 million year old *Taquatomys decorata* of North Africa. These species all had very narrow jaw ridges, the shell lacks mesoplastra and the first costals reach the shell margin.

The next family to be found in the fossil record is the Euraxemydidae, comprising only *Euraxemys essweini* from the Santana Formation of Brazil and *Dirqadim schaefferi* from Morocco. These both date from 100 million years ago, having much the same distribution as the Araripemydidae. They are unusual turtles in that the quadrate bone of the skull expands medially and covers the prootic bone and makes a narrow contact with a ventral process of the exoccipital. In all other pleurodires the prootic is completely exposed or completely covered ventrally and the exoccipital reaches the base of the skull, extending into an elongate foot next to the basioccipital. This family seems to be related to the superfamily Podocnemidoidea which today include the South American and Madagascan river turtles (Podocnemididae).

Of the Euraxemydidae-Podocnemididae lineage the earliest family was the Bothremydidae. This was a diverse family of animals with very robust skulls. They had a wide contact at the back of the skull between the exoccipital and quadrate bones, a Eustachian tube separated from the ear canal, which forms a bony canal for the stapes, very little cheek emargination with contact between the supraoccipital and quadrate bones (except in some species) and between the maxilla and quadratojugal, and a posterior enlargement of the eye socket. The family extended from 112 million years

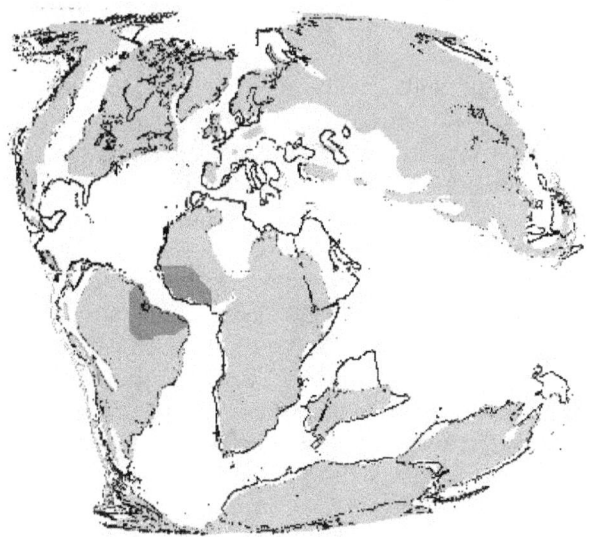

Origins of podocnemidods from 110 million years ago.

ago to 50 million years ago and were found throughout the Americas, Europe, Africa, and India. Although they seem to have evolved first in the South America – North Africa area of Gondwana they spread into the other continents by crossing relatively wide expanses of ocean. Their movement northwards may only have required the crossing of narrow, shallow stretches of water, but they would have had to cross open ocean to reach India which had separated from Africa prior to the origin of the family, and would have had a more significant oceanic crossing to North America and Europe. If they spread across what remained of Gondwana shortly after the first appearance of fossil bothremydids 112 million years ago they could have crossed into India-Madagascar across the relatively narrow Mozambique Channel. A similarly narrow channel would have enabled them to cross from North Africa into Europe and thence to North America. However, there is a considerable gap between this timing and the appearance of fossil bothridemydis in Madagascar (*Kinkonychelys* 70 million years ago) and India (*Sankuchemys* and *Kurmademys* 65 million years ago). These are contemporaneous with the European and North American bothridemydids (70 million years ago), these dates may suggest that the bothridemydids did not spread until around 70 million years ago when great ocean crossings would be needed. Alternatively these discrepancies may however be simply the result of a patchy fossil record.

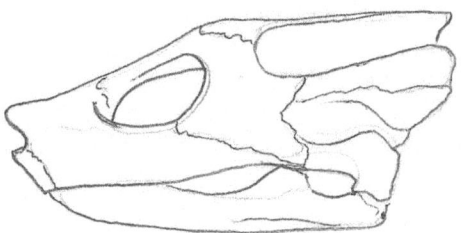

Bothridemydid skull

The bothridemydids were followed by a range of South American forms. The first of these were the genera *Brasilemys*, *Portezueloemys* and *Hamadachelys* from Argentina and Brazil around 100 million years ago. These in turn gave rise to the related subfamilies Bauremydinae and Podocnemidinae. Bauremydinae were also restricted to South America, with the 80 million year old genus *Bauremys*.

Although there is one living podocnemidine outside of South America this subfamily seems to have originated in that continent. The first representatives were the 80 million year old group of *Peiropemys, Pricemys* and *Lapparentemys*. At this time South America was isolated from other continents but the separation from Africa was by a relatively narrow sea-way; as with the bothridemydids, this was not a barrier for the movement of these large aquatic turtles. The early podocnemidines were followed by two groups, one forming the living *Podocnemis* and the 15 million year old *Caninemys*, the other forming a diverse group of largely African genera. These seem to have been present on the continent from at least 80 million years ago as indicated by some hard to identify fossils, although good fossil material only dates from 50 million years ago.

54

They seem to have colonised North Africa from South America and dispersed in two directions; south into East Africa and Madagascar, and north-eastwards. The early species include the 35 million year old Egyptian *Dacquemys*, followed by the southern group of the Kenyan *Turkanemys* (5 million years ago) and the living Madagascan *Erymnochelys* (fossils of which may date back as far as 65 million years ago). The northwards dispersal led to the 50 million years old marine European and North African *Neochelys*, followed by the French *Papoulemys* and a group that dispersed eastwards. This easterly dispersal starts with other Egyptian species, *Cordichelys, Latenemys, Lemurchelys* and *Mogharemys*, giving rise to *Brontochelys* from Pakistan and *Shweboemys* from Burma, all ranging from 35 to 15 million years ago. Many of these were marine forms, including *Neochelys* which was found on both sides of the Tethys seaway. The two exceptions to this African and Eurasian grouping are the living South American *Peltocephalus* which is apparently closely related to the Madagascan *Erymnochelys* and the 5 million year old Venezuelan *Bairdemys* which is close to the Egyptian *Latenemys*. At first sight these would appear to suggest that podocnemidinines moved back across the Atlantic more recently. Whilst this is not impossible it seems to run counter to the general trend in the subfamily. Whether or not *Bairdemys* is really a relative of the Egyptian *Latenemys* remains uncertain. *Peltocephalus* has probably not colonised South America from Africa or Madagascar but is probably a relict of to the ancestral South American form of the *Erymnochelys* lineage from 80 million years ago.

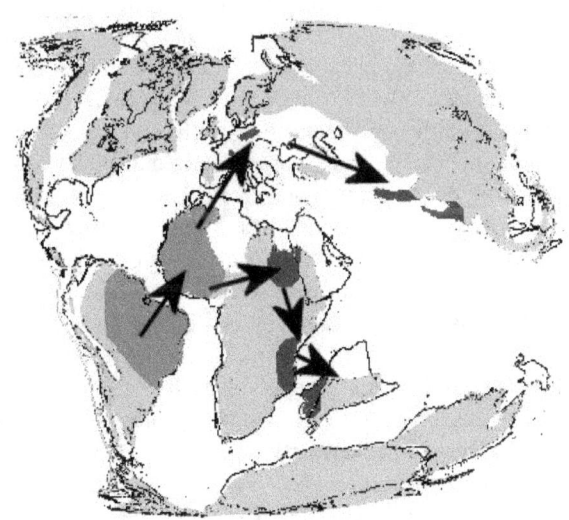

Map of podocnemididine dispersal from 80 million years ago. Later expansion shown in dark grey.

Today *Peltocephalus* comprises a single species: the big-headed Amazon river turtle *P. dumerilianus* of the Amazon, from Brazil to Venezuela and Colombia. This is a denizen of large forest rivers where it feeds on plants and animals. It seems to be the most carnivorous of the South American

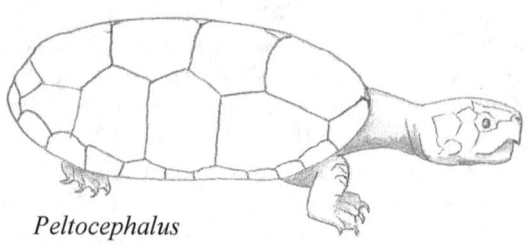

Peltocephalus

podocnemidids, consuming a high proportion of fish. In its appearance and in its diet it is similar to the related Madagascan *Erymnochelys*. *Erymnochelys* is the only surviving non-South American podocnemidine, the Madagascar big-headed turtle or ré-ré *E. madagascariensis*. This species is found only in western Madagascar, where it occurs in lakes and large rivers. It is a large and powerful freshwater turtle, feeding on fish and amphibians, but also taking birds occasionally. Reproduction in this species is rapid, with up to 60 eggs being laid in a year and maturity being attained as early as 5 years old. Despite this the

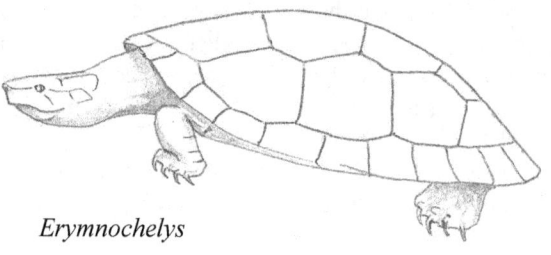

Erymnochelys

species is threatened by human consumption and habitat loss.

In addition to these reasonably well known groups, there are a number of fragmentary fossils that belong to genera of unknown affinities: *Apoidochelys*, *Cearachelys*, *Palaeaspis* (or *Palaeochelys* or *Palemys*), *Paralichelys*, *Potamochelys* and *Stupendemys*. *Stupendemys geographicus* was a giant turtle from around 10 million years ago in Venezuela and Brazil estimated to have been 3.3 m long and 2.18 m wide. This massive animal is known only from some shells and leg bones so its relationship to other podocnemids is not clear. More recent fossil South American turtles have mostly been assumed to belong to the genus *Podocnemis* but they are very poorly defined.

The *Podocnemis* lineage includes the fossils *Bauremys*, *Roxochelys* (doubtful identification) and *Brasilemys*. *Brasilemys josai* is from Brazil 65 million years ago, it is particularly notable as it seems to be a well defined ancestor of living *Podocnemis* as well as of the fossils *Hamadachelys escuilliei* and the Colombian *Cerrejonemys wayuunaiki* (doubtful identification). Within *Podocnemis* there are six species: the red-headed river turtle *P. erythrocephala*, Arrau sideneck *P. expansa*, Magdalena river turtle *P. lewyana*, six-tubercled river turtle *P. sextuberculata*, yellow-spotted river turtle *P. unifilis* and the savannah side-necked turtle *P. vogli*.

The red-headed river turtle is a very distinctive species. It is found only in slow moving waters of the Rio Negro of Venezuela and Colombia. Adults are mainly vegetarian, often feeding on surface debris by suction. The Arrau sideneck is a widespread species in north-western Brazil to Bolivia and Guyana in the Amazon, Orinoco and Essequibo river systems. It is the largest pleurodire and may reach 90 kg in weight, with a length of 90 cm. Its powerful build allows it to occupy all the larger rivers and to migrate over land between rivers. It is locally abundant and in the past used to form basking aggregations of thousands of individuals. The Magdalena river turtle is found in the Magdalena and Sinu rivers of Colombia. Despite its disjunct distribution it is genetically highly uniform.

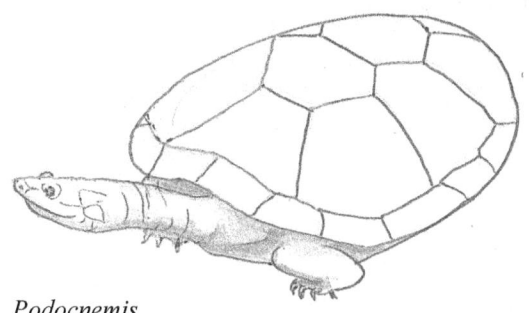

Podocnemis

This genetic uniformity and lack of any geographical pattern suggests that the now widely separated populations used to move freely but have recently passed through a population bottleneck. The six-tubercled river turtle is a predatory species from large rivers of the Amazon Basin of Brazil, Peru and Colombia. The yellow-spotted river turtle is found throughout the Amazon and Orinoco basins of South America. It is mainly found in smaller, quiet rivers and is largely herbivorous. The savannah side-neck is found in the Orinoco floodplain of Venezuela and Colombia where large numbers (sometimes thousands) may congregate in pools of water. It is an opportunistic feeder.

The relationships between these species are not clear. The most plausible evolutionary scenario is that the ancestral form was a widespread generalist resembling the Arrau sideneck. This gave rise to the large powerful species of the main rivers, the Arrau sideneck, and smaller species of quieter waters. The smaller species divide geographically into a northern form (the savannah sideneck of the Orinoco) and a group centred on the main Amazon River. The six-tubercled retains the Amazon distribution with more back-water forms evolving in the main range (yellow-spotted), localised in the Rio Negro (red-headed) and isolated in the west (Magdalena river turtle). These species originated around 30 million years ago. At this time climatic change and geological upheaval caused by the formation of the Andes Mountains would have changed river flows; in particular a cooler, dryer climate would have isolated many river animals, leading to speciation in newly isolated river systems. These South American river turtles seem to show a mixture of speciation by isolation as expected, but also specialisation into main river and back-water forms.

Pelomedusidae are known from 120 million years ago. There are two living genera: *Pelomedusa* and *Pelusios*. *Pelomedusa* are large terrapins found throughout Africa. *Pelusios* are much smaller and have a hinged plastron, being able to draw up the anterior lobe to close the front opening of the shell in all but one species. Although soft-shelled turtles are present in the larger water-bodies of Africa *Pelusios* and *Pelomedusa* are the dominant aquatic turtles of the continent. Emydine terrapins do occur along the northern fringe of the Sahara but they only reached the continent relatively recently and have not been able to cross the arid barrier of the desert.

Pelomedusa fossils date from around 19 million years ago with the species *P. senutpickfordina* from Namibia. The only living member of the genus, the helmeted terrapin *P. subrufa* is the most widespread species of non-marine turtle. Some subspecies have been described but their validity is questionable. DNA identifies three well-defined groups. These

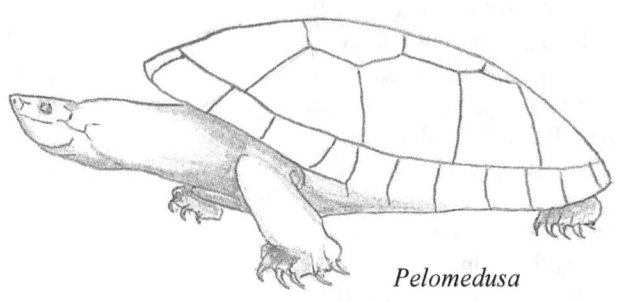

Pelomedusa

are the northern African group (Ghana to Kenya), a north-eastern group (Somalia to Arabia) and a southern group (from the Democratic Republic of Congo southwards, including Madagascar). The northern group shows some differences between samples from Cameroon, those from Ghana and Ivory Coast, those from Benin, Burkina Faso and Niger, those from the Central African Republic, and those from Kenya). The north-eastern group has differences between samples from Somalia, and from Arabia. The southern group has differences between most populations and the South African samples. The Madagascar population is identical to that of the main southern African grouping and seems to be recently arrived on the island. At present there are no estimates of how old these three genetic groups are, but they probably date back to the last series of Ice Ages when alternating periods of ice expansion and retreat towards the poles was mirrored in Africa by periods of wet and dry conditions. During dry periods the helmeted terrapins were probably isolated in the wet Congo basin (the southern genetic form), a forest refuge in west Africa (the northern form) and somewhere near the shores of the Red Sea (the north-eastern form).

    Pelusios fossils date from 18 million years ago with *P. rusingae* from Kenya. Fossils supposedly identifiable as the living species *P. sinuatus* have been recorded

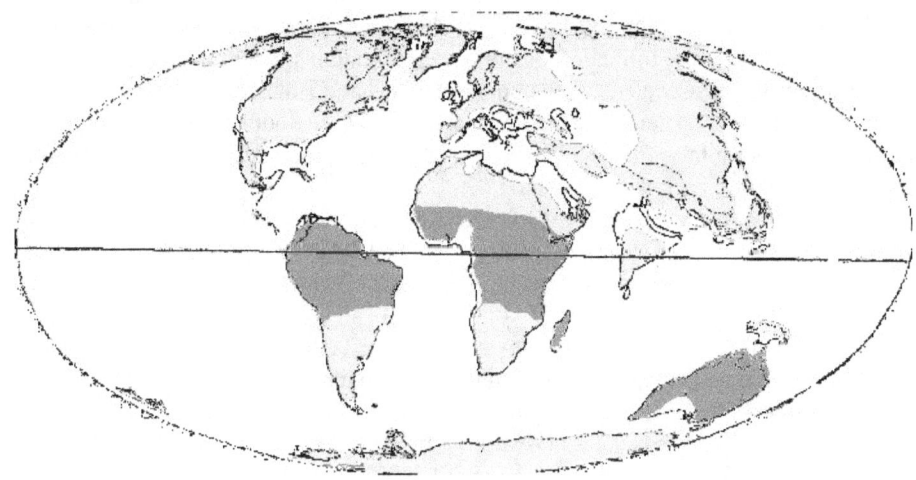

Pelomedusidae 50 million years ago; the family is now restricted to Africa.

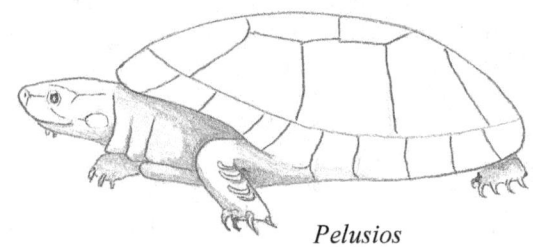

Pelusios

from Kenya and Chad although the species now is restricted to the south-east of Africa, these are very poor fossils and probably not really identifiable as *P. sinuatus*. However, they are almost certainly a *Pelusios* of some sort. *Pelusios* has been divided into two morphological groupings: the *subniger* and *castaneus* groups. However, these are clearly unrealistic and more recent morphological groupings consider both *subniger* and *castaneus* to be members of one major branch of the *Pelusios* tree. Recent analysis suggests that the most isolated species is the Gabon mud turtle *P. marani*. The west African range of this species may be close to the original home of the genus, and is still the area with the greatest diversity of *Pelusios* forms. West African *Pelusios* appear to have spread rapidly south-eastwards, giving rise to the south-eastern serrate hinged turtle *P. sinuatus* and the *P. subniger* group. In addition to these south-eastern species the main radiation of *Pelusios* species probably remained in the Congo rainforest of west Africa.

The *subniger* group (*subniger, upembae, bechuanicus*) is southern-eastern African, ranging from the south-eastern East African black mud turtle *P. subniger*, the inland eastern Upemba mud turtle *P. upembae*, and the Okavango mud turtle *P. bechuanicus*.

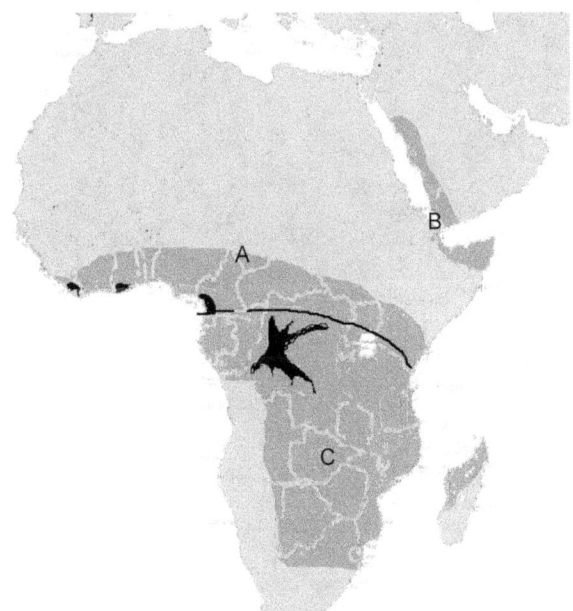

Approximate distribution of *Pelomedusa* genetic groups and the main Ice Age refugia in Africa (black).

The main *Pelusios* radiation comprises several west African species with narrow ranges and some more wide ranging forms. They can be divided into the African dwarf mud turtle *P. nanus* from the southern edge of the Congo, the western Congo species pair of the Ivory Coast mud turtle *P. cupulatta* and the more inland West African black mud turtle *P. niger*, and a complex that originates in the central Congo. This complex comprises the African forest turtle *P. gabonensis* and the related *castaneus* and *castanoides* groups.

The *castaneus* group probably originated in the central Congo, splitting into two pairs of species, an eastern and a central-western pair. The eastern species are Adanson's *P. adansoni* (north-east end of the Congo) and the Turkana mud turtle *P. broadleyi* (east-central Africa). The Turkana mud turtle is the only species of *Pelusios* without a hinged plastron. The central-western pair is the Central African mud turtle *P. chapini* (Congo river) and the West African mud turtle *P. castaneus* which may have moved westwards from an ancestrally more central range. These two species are genetically difficult to separate and differ largely on size, so may not really be distinct. The extinct Seychelles terrapin *P. seychellensis* is generally thought to be related to the West African mud turtle, but this is hard to explain on biogeographical grounds. The West African mud turtle appears to contain a moderate level of genetic diversity and may still include some undescribed species; there appears to be a division between the populations introduced

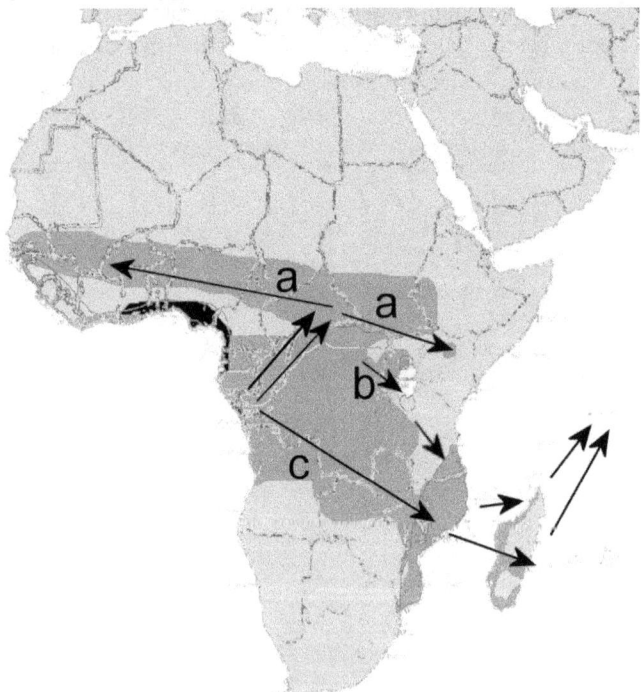

Dispersal of the *Pelusios* groups from the Congo: a – *castaneus* group; b – *castanoides* group; c – *subniger* group; possible ancestral range shown in black.

to Guadeloupe and the Congo-Nigeria populations. Within the latter there is a notable degree of separation between the Congo population and the West African samples. The population of Sao Tome seems to be descended from animals from Ivory Coast.

The *castanoides* group comprises the south-eastern African yellow-bellied mud turtle *P. castanoides* and the southern Congo pairing of the Mashona hinged terrapin *P. rhodesianus* and the African keeled mud turtle *P. carinatus*. Williams' mud turtle *P. williamsi* probably belongs in this group but this in not clear at present. Williams' contains the Lake Victoria mud turtle *williamsi*, Ukerewe island mud turtle *laurenti* and the Albertine Nile mud turtle *lutescens*. It seems that this group originated in the southern Congo (Mashona and African keeled mud turtles), spreading eastwards (Williams' and eventually the yellow-bellied). There is some mismatch between morphological and genetic identification in this group, with some confusion between Mashona and African keeled mud turtles. Specifically it seems that the earliest genetic form is the Angolan population of *rhodesianus*, followed by the Burundi *rhodesianus* which groups with *carinatus*. This may suggest either a misidentification from Burundi or that the group originates in the south, spreading northwards.

Where the ranges of *Pelusios* terrapins overlap their precise distribution may be determined by unexpected factors. For example, in Tanzania water bodies contain large number or serrated hinged terrapin or East African black mud turtles, but rarely similar numbers of each. This is not because of competition between them but rather is the effect of the presence of crocodiles in some areas. The small, flattened shells of the black mud turtle are easily crushed by Nile crocodiles, and make up the commonest prey item for crocodiles after fish, but the serrated hinged terrapin is able to live alongside crocodiles because its high domed and keeled shell is able to withstand crushing much better.

The cause of the isolation that led to the evolution of the different African mud turtle species is not known but can probably be attributed to the Ice Age effects that have structured helmeted terrapin populations. For much of the period from 22,000 to 11,000 years ago dry conditions prevailed and the present day extensive rainforests of West Africa and the Congo were restricted to isolated blocks of forest along the West African coast and in the centre of the Congo basin. At this time forest turtles would have been isolated in three or four areas and may have started to diverge into the present species.

Very little is known of the true diversity of *Pelusios* species and their evolutionary history, this genus having been largely neglected by biologists. The origins of the populations are difficult to determine due to a lack of research in much of the west African range. In the east there is less diversity and the distributions better known. The most easterly populations are in the Seychelles islands where there appear to have been two waves of colonisation, the earliest being the Seychelles terrapin which resembles the West African mud turtle but is thought to be distinct. More recent colonisation may have given rise to endemic subspecies of the East African black mud turtle (*P. subniger parietalis*) and the yellow-bellied mud turtle (*P. castanoides intergularis*). Both seem to have a recent origin, either during periods of low sea level in the most recent Ice Ages over the past 5 million years, or as a result of human introduction in the past two centuries. For the yellow-bellied mud turtle the Seychelles population is very closely related to the Madagascan '*P. castanoides kapika*', and less closely to the East African population. This suggests movement from Africa to Madagascar, and then to Seychelles. Relationships within the East African black mud turtle have not been studied. The Seychelles populations are in rapid decline, the two endemic subspecies being critically endangered and the Seychelles terrapin being extinct. This last species is known from only three specimens collected in 1895.

References

Cadena Rueda, E.A. & E.S. Gaffney. 2005. *Notoemys zapatocaensis*, a New Side-Necked Turtle (Pleurodira: Platychelyidae) from the Early Cretaceous of Colombia. *Amer.Mus.Novit.***3470**: 1-19

Cadena, E., C. Jaramill & M.E. Paramo. 2008. New material of *Chelus colombiana* (Testudines; Pleurodira) from the Lower Miocene of Colombia. *Journ. Vert. Palaeont.* **28**(4): 1206-1212

Caro, T. & H.B. Shaffer. 2010. Chelonian antipredator strategies: preliminary and comparative data from Tanzania *Pelusios. Chel. Conserv. Biol.* **9**: 302-305

Fritz, U., W.R. Branch, M.D. Hofmeyr, J. Maran, H. Prokop, A. Schleicher, P. Siroky, H. Stuckas, M. Vargas-Ramirez, M. Vences & A.K. Hundsdorfer. 2010. Molecular phylogeny of African hinged and helmeted terrapins (Testudines: Pelomedusidae: *Pelusios* and *Pelomedusa*). *Zool. Script.* **40**: 115-125

Fuente, M.S. de la. 2003. Two new pleurodiran turtles from the Portezuelo Formation (Upper Cretaceous) of Northern Patagonia, Argentina. *Journ. Paleontol.* **77**: 559-575

Fuente, M. de la, F. de Lapparent de Broin & T. Manera de Bianco. 2001. The oldest and first nearly complete skeleton of a chelid, of the Hydromedusa sub-group (Chelidae, Pleurodira), from the Upper Cretaceous of Patagonia. *Bull. Soc. Geol. Fr.* **172**: 237-244

Gaffney, E.S., D. de A. Campos & R. Hirayama. 2001. *Cearachelys*, a New Side-Necked Turtle (Pelomedusoides: Bothremydidae) from the Early Cretaceous of Brazil. *Amer. Mus. Novitates* **3319**: 20

Gaffney, E.S. & C.A. Forster. 2003. Side-Necked Turtle Lower Jaws (Podocnemididae, Bothremydidae) from the Late Cretaceous Maevarano Formation of Madagascar. *Amer. Mus. Novit.* **3397**

Gaffney, E.S., D.W. Krause & I.S. Zalmout. 2009. *Kinkonychelys*, A New Side-Necked Turtle (Pelomedusoides: Bothremydidae) from the Late Cretaceous of Madagascar. *Am. Mus. Novitates* **3662**: 1-25

Gaffney, E.S., P.A. Meylan, R.C. Wood, E. Simons & D. De Almeida Campos. 2011. Evolution of the side-necked turtles: the family Podocnemididae. *Bull. AMNH* **350**: 1-237

Gaffney, E.S., H. Tong & P.A. Meylan. 2006. Evolution of the side-necked turtles: the families Bothremydidae, Euraxemydidae, and Araripemydidae. *Bull. AMNH* **300**: 1-698.

Gallo de Franca, M.A. & M. Cardoso Langer. 2006. Phylogenetic relationships of the Bauru group turtles (Late Cretaceous of South-Central Brazil). *Rev. bras.*

paleontol. **9**(3):365-373

Gallo de Franca, M.A. & M.C. Langer. 2006. Phylogenetic relationships of the Bauru Group turtle (Late Cretaceous of South-Central Brazil). *Rev. bras. paleontol.* **9**(3)

Gasparini, Z. & M. Fernández. 2005. Jurassic marine reptiles of the Neuquén Basin: records, faunas and their palaeobiogeographic significance. *Geological Society, London, Special Publications* **252**: 279-294

Georges, A. & S. Thomson. 2010. Diversity of Australasian freshwater turtles, with an annotated synonymy and keys to species. *Zootaxa* 2496

Giles, J.C., J.A. David, R.D. McCauley & G. Kuchling. 2009. Voice of the turtle: The underwater acoustic repertoire of the long-necked freshwater turtle, *Chelodina oblonga. J. Acoust. Soc. Am.* **126**: 434-443

Iverson, J.B., R.M. Brown, T.S. Akre, T.J. Near, M. Le, R.C. Thomson & D.E. Starkey. 2007. In Search of the Tree of Life for Turtles. *Chel. Res. Monogr.* **4**: 85–106

Kischlat, E.-E. 1993. *Quélidas (Chelonii, Pleurodira) da Bacia de Taubaté, Cenozóico do Estad de Sap Paulo, Brasill.* Unpublished MSc thesis, Universidade Federal do Rio de Janeiro.

Kuchling, G. 1999. An update of the Western Swamp Tortoise *Pseudemydura umbrina* captive breeding and reintroduction programme. *Testudo* **5**(1): 17-22

Lapparent de Broin, F. 2000. African chelonians from the Jurassic to the present: phases of development and preliminary catalogue of the fossil record. *Paleontologia Africana* **36**:43–82.

Lapparent de Broin, F. 2003. *Neochelys* sp. (Chelonii, Erymnochelyinae), from Silveirinha, early Eocene, Portugal. *Ciências da Terra* **15**

Lapparent de Broin, F. de, X.M. Bereikua & V. Codrea. 2004. Presence of Dortokidae (Chelonii, Pleurodira) in the earliest Tertiary of the Jibou Formation, Romania: Paleobiogeographical implications. *Acta Palaeontol. Rom.* **4**: 203-215

Lapparent de Broin, F. de & M.S. de la Fuente. 2001. Oldest world Chelidae (Chelonii, Pleurodira), from the Cretaceous of Patagonia, Argentina. Comptes Rendus de l'Académie des Sciences. IIA: *Earth Planet. Sci.* **333**: 463-470

Lapparent de Broin, F., M.S. de la Fuente & M.S. Fernandez. 2007. *Notoemys laticentralis* (Chelonii, Pleurodira), Late Jurassic of Argentina: new examination of the anatomical structures and comparisons. *Rev. Paléobiol.* **26**(1): 99-136

McCord, W.P., M. Joseph-Ouni & W.W. Lamar. 2001. Taxonomic Reevaluation of *Phrynops* (Testudines: Chelidae) with the description of two new genera and a new species of *Batrachemys. Rev. biol. Trop.* **49**(2)

Megirian, D. & P. Murray. 1999. Chelid turtles (Pleurodira, Chelidae) from the Miocene Camfield Beds, Northern Territory of Australia, with a description of a new genus and species. *The Beagle* **15**: 75-130

Meylan, P.A. 1996. Skeletal morphology and relationships of the Early Cretaceous side-necked turtle, *Araripemys barretoi* (Testudines: Pelomedusoides. *Journ. Vert. Palaeont.* **16**: 20-33

Restrepo, A., B.C. Bock & V.P. Páez. 2008. Genetic variability in the Magdalena River turtle, *Podocnemis lewyana* (Duméril, 1852), in the Mompos Depression, colombia. *Actu Biol.* **30**

Scheyer, T.M., B. Brüllmann & M.R. Sánchez-Villagra. 2008. The ontogeny of the shell in side-necked turtles, with emphasis on the homologies of costal and neural bones. *Journ. Morphol.* **269**(8): 1008-1021.

Seddon, J.M., A. Georges, P.R. Baverstock & W. McCord. 1997. Phylogenetic Relationships of Chelid Turtles (Pleurodira: Chelidae) Based on Mitochondrial 12S rRNA Gene Sequence Variation. *Mol. Phyl. Evol.* **7**: 55-61

Souza, F.L., A.F. Cunha, M.A. Oliveira, G.A.G. Pereira & S.F. dos Reis. 2003. Preliminary Phylogeographic Analysis of the Neotropical Freshwater Turtle *Hydromedusa maximiliani* (Chelidae). *Journ. Herpetol.* **37**: 427-433

Thomson, S., A. White & A, Georges, 1999. Re-evaluation of *Emydura lavarackorum*: identification of a living fossil. *Mem. Queensl. Mus.* **42**: 327-336

Thomson, S.A. 2000. A revision of the fossil chelid turtles (Pleurodira) described by C.W. de Vis, 1897. *Mem. Queensl. Mus.* **45**(2): 593-598.

Thomson, S.A. & B.S. Mackness. 1999. Fossil Turtles from the Early Pliocene Bluff Downs Local Fauna, with a description of a new species of *Elseya*. *Trans. R. Soc. S. Aust.* **123**(3): 101-105

Vargas-Ramírez, M., O.V. Castaño-Mora & U. Fritz 2008. Molecular phylogeny and divergence times of ancient South American and Malagasy river turtles (Testudines: Pleurodira: Podocnemididae). *Organisms Diversity & Evolution* **8**: 388-398.

Vargas-Ramírez, M., M. Vences, W.R. Branch, S.R. Daniels, F. Glaw, M.D. Hofmeyer, G. Kuchling, J. Maran, T.J. Papenfuss, P. Iroký, D.R. Vieites & U. Fritz. 2010. Deep genealogical lineages in the widely distributed African helmeted terrapin: Evidence from mitochondrial and nuclear DNA (Testudines: Pelomedusidae: *Pelomedusa subrufa*). *Mol. Phyl. Evol.* **56**: 428-440

6. The early cryptodires

Cryptodires comprised two major groups, the Paracryptodires and the Eucryptodires. The latter include all the living cryptodires and fit the conventional understanding of this group. Paracryptodires are difficult to make sense of; they may be ancestral cryptodires or a completely separate group that just resemble cryptodires in that they are not pleurodires.

Paracryptodires contain two distinct families, Baenidae and Pleurosternidae, and one genus of uncertain affinities (*Dorsetochelys*). There were several Baenidae genera from North America from 100 to 45 million years ago. The earliest forms *Trinitichelys* and *Neurankylus* showed well developed temporal emargination and thin posterior margins of the parietal bones. These were followed by two groups of more advanced species, showing less emargination and thicker parietals: one group is initially characterised by *Plesiobaena*, followed by the more advanced North American grouping from 65 to 56 million years ago: *Peckemys*, then *Cedrobaena*, and *Gamerabaena* and *Palatobaena*. The other group contains the pairing of *Boremys* and *Eubaena*, and the group of *Gloremys*, *Stygiochelys* and the pair of *Baena* and *Chisternon*. The trend of strengthening of the skull seen in the baenids may be a defensive response to the increasingly efficient mammalian predators around at this time. This selection for strengthening may also account for the unusual degree of shell fusion found in adult baenids; although turtle growth slows down with age most species retain the ability to grow throughout their lives, but a few species (and all baenids) have fully fused shells, making them incapable of growth in adulthood.

Pleurosternidae include *Chengyuchelys*, *Compsemys*, *Desmemys*, *Dinochelys*, *Glyptops*, *Mesochelys*, *Pleurosternon* and *Uluops* from North America and Europe. At least some species (e.g. *Compsemys*) are thought to have inhabited stationary or very slow moving water, rather than the river habitat occupied by the Baenidae. The poorly defined Solemydidae (*Helochelydra*, *Naomichelys* and *Solemys*) may also be related to these families. Solemydidae were Laurasian, found in North America and the west of Europe from 145 to 65 million years ago. *Solemys* was the last surviving genus, it was restricted to the Iberian Peninsula and southern France from about 80 to 65 million years ago. It may have been terrestrial but was probably semi-aquatic in brackish water environments.

The Eucryptodira, or 'true' cryptodires have an enclosed internal carotid artery, the canal for which extends to the back of the pterygoid bones in the palate. The early species were distributed across Asia (mainly in the Xinjiangchelydae) with a few poorly defined and largely un-named species in Europe and North Africa. Marine Eucryptodires are also found in fossil deposits from North and South America from 100 million years ago to around 65 million years ago, and some were recorded in Australia. The ancient marine turtles (not related to living marine turtles) comprise the early *Thalassemys moseri* and *Santanachelys gaffneyi* (these two species being related), *Portlandemys* and *Plesiochelys*, and more advanced species. *Plesiochelys* and some other genera were thought to form a distinct family, the Plesiochelyidae, but this family now seems

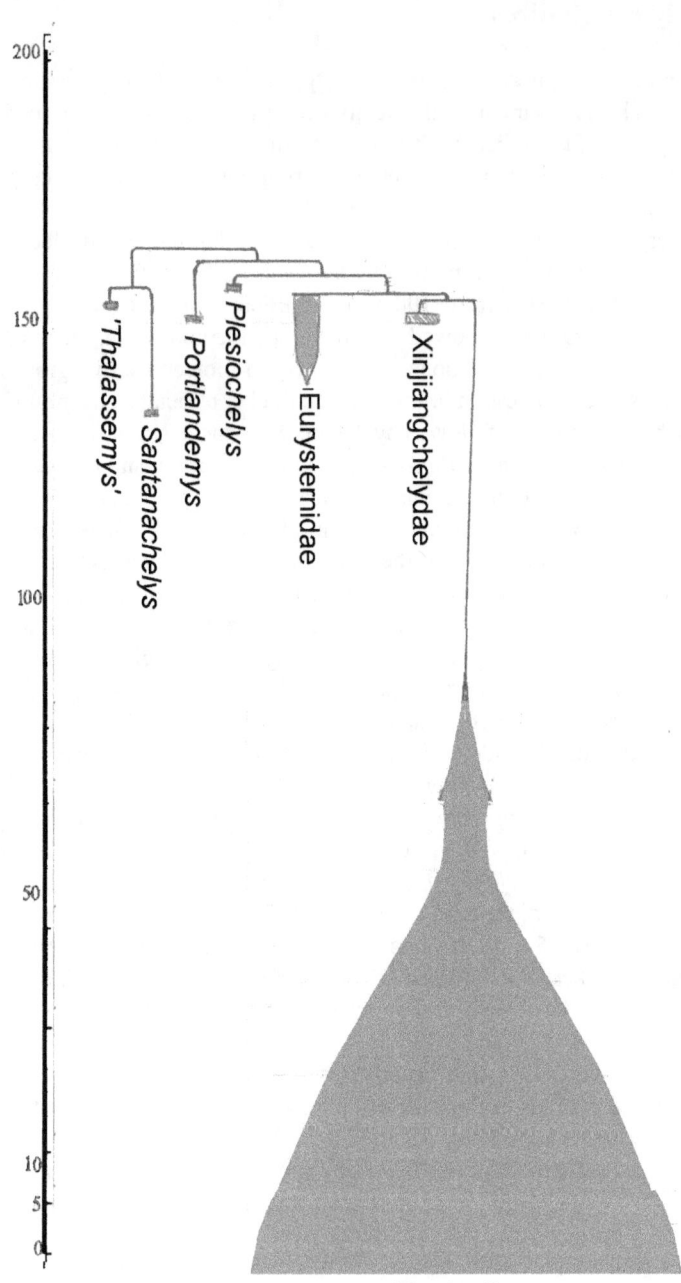

Eucryptodire phylogeny

Solnhofia skull (ventral) and *Angolachelys* skull in lateral view and reconstruction

to be just a collection of early eucryptodires. The more advanced forms comprise the Eurysternidae *Solnhofia, Sandowia* and *Angolachelys*. The eurysternids were still large marine turtles but were related to the freshwater Xinjianchelydae and Testudines. These groups all originated between 150 and 120 million years ago.

The fossil family of largely Chinese turtles, the Xinjiangchelydae, comprise some well preserved fossils, particularly *Xinjiangchelys qiguensis* from 145 million years ago of north-west China. These appear to be eucryptodires, having lost the mesoplastral bones. Xinjiangchelyds were some 40 cm long, with a flattened shell. These were known from possibly as early as 200 million years ago from Sichuan, Xinjiang, Kirghizia and the Transaltai Gobi of Mongolia. '*Plesiochelys*' *chungkingensis* is a xinjiangchelyd with a typical carapace but showing the primitive *Plesiochelys* plastral condition of retaining mesoplastra.

The relatively well defined *Xinjianchelys* species are found in central Asia (Kazakstan to Mongolia and north-west China). Related to these were *Chengyuchelys* and *Sichuanchelys* in east-central Asia (central China) and *Siamochelys peninsularis* in Thailand. *Chengyuchelys* retains mesoplastra (like early turtles) but these have been lost in several of the earlier pancryptodires and are lost repeatedly in later forms. These xinjiangchelyids seem to have been restricted to the freshwater habitats of east Laurasia and, although successful for at least 55 million years, they probably did not leave any descendants after 145 million years ago. The Xinjianchelydae appear to have been freshwater and the transition from dry to more humid conditions 145 million years ago may have been expected to benefit this family. However the favourable climatic conditions were also accompanied by high sea levels which may have flooded low-lying rivers and marshes occupied by the xinjianchelyds. The earliest Testudines seem to have been marine and estuarine, and so would not have been affected by this change.

Testudines (tortoises and terrapins) comprise several species from about 120 million years ago when they are more accurately labelled Paracryptodira. These include the Chinese and Mongolian *Hangaiemys, Kirgizemys, Dracochelys, Sinemys* and *Ordosemys*, and the North American *Judithemys*. *Sinemys* may be related to the poorly known Australian *Chelycharapookus arcuartus*, forming the Sinemydidae. These animals may have been marine, which would explain how Sinemydidae could be found in Asia and Australia, separated by hundreds of kilometres of ocean. Some of these fossils may belong to the

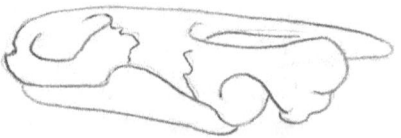

skull of *Judithemys*

Macrobaenidae but their identification is far from certain. In addition to these extinct families, paracryptodires include the living cryptodires. These living cryptodires form three main groups: the soft-shelled turtles (Trionychia); a group comprising the snapping turtles, mud turtles and sea turtles; and the Testudinoidea (tortoises and terrapins). Although these are widely used groupings, DNA studies struggle to present well supported patterns of relationships, suggesting that most main branches of cryptodires evolved in a rapid radiation between 120 and 90 million years ago.

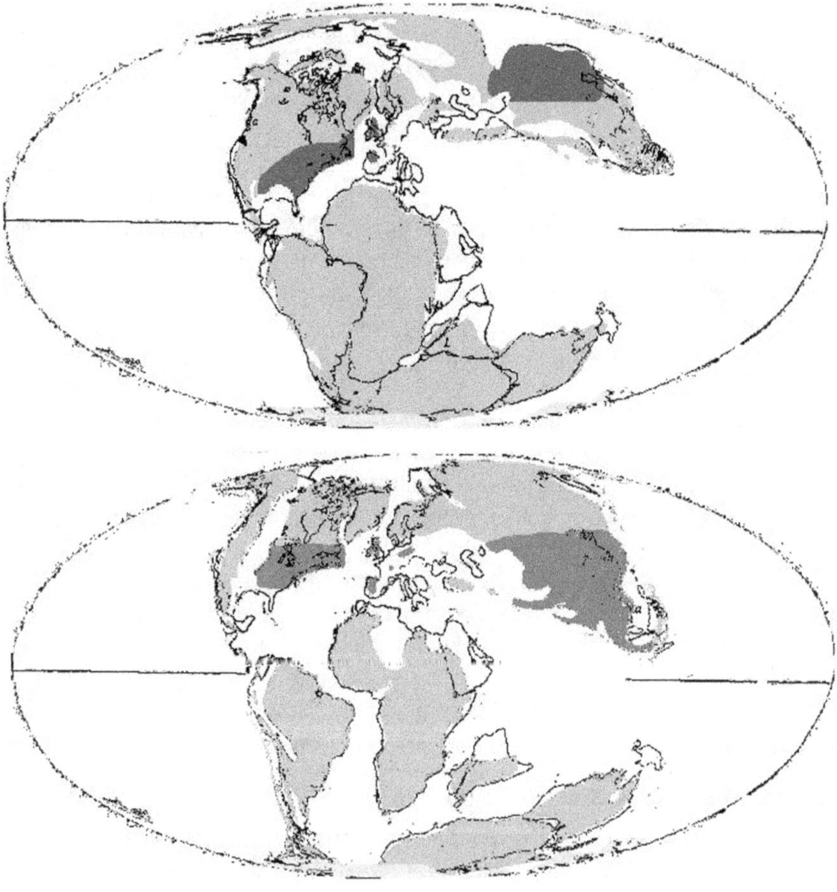

Non-marine cryptodire distribution 150 and 100 million years ago.

References

Danilov, I.G. & J.F. Parham. 2008. A reassessment of some poorly known turtles from the Middle Jurassic of China. *Journ. Vert. Paleont.* **28**(2): 306-318

Datta, P.M., P. Manna, S.C. Ghosh & D.P. Das. 2000. The first Jurassic turtle from India. *Palaeontology* **43**: 99-103

Gaffney, E.S. 1996. The postcranial morphology of *Meiolania platyceps* and a review of the Meiolaniidae. *Bull. AMNH* **229**

Gaffney, E.S., J.H. Hutchison, F.A. Jenkins & L.J. Meeker. 1987. Modern turtle origins: The oldest known cryptodire. *Science* **237**: 289-291

Gaffney, E.S., L. Kool, D.B. Brinkman, T.H. Rich & P. Vickers-Rich. 1998. *Otwayemys*, a new cryptodiran turtle. *Amer.Mus.Novit.***3233**

Gaffney, E.S., T.H. Rich, P. Vickers-Rich, A. Constantine, R. Vacca & L. Kool. 2007. *Chubutemys*, and the relationships of the Meiolaniidae. *Amer.Mus.Novit.***3599**

Hirayama, R., D.B. Brinkman & I.G. Danilov. 2000. Distribution and biogeography of non-marine Cretaceous turtles. *Russ. Journ. Herpetol.* **7**: 181-198

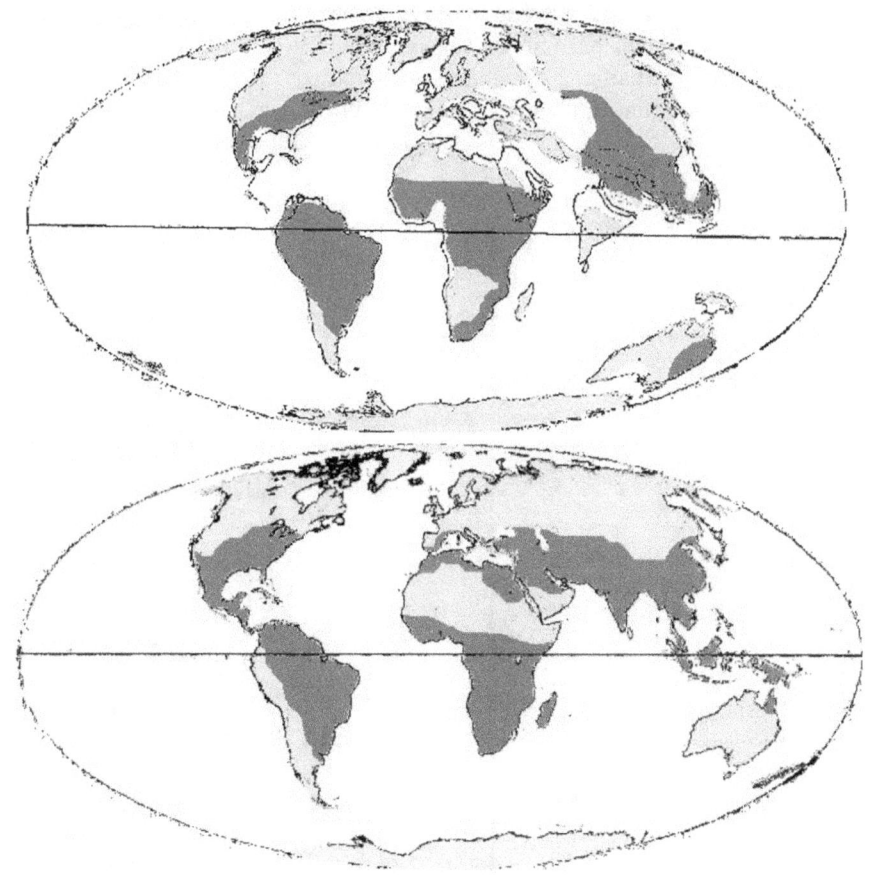

Non-marine cryptodire distribution 50 million years ago and at the present day.

Hutchison, J.H. 1984. Determinate growth in the Baenidae (Testudines): taxonomic ecologic, and stratigraphic significance. *Journ. Vert. Palaeont.* **3**: 148-151

Lipla, T.R., F. Therrien, D.B. Weishampel, H.A. Jamniczky, et al. 2006. A new turtle from the Arundel Clay. *Journ. Vert. Paleontol.* **26**: 300-307

Lyson, T.R. & W.G. Joyce. 2009. A new species of *Palatobaena* and a phylogenetic analysis of Baenidae. *J. Palaeont.* **83**: 457 & A revision of *Plesiobaena* and an assessment of Baenid ecology. *J. Palaeont.* **83**: 833-853

Lyson, T.R. & W.G. Joyce. 2011. Cranial anatomy and phylogenetic placement of the enigmatic turtle *Compsemys victa* Leidy, 1856. *J. Palaeont.* **85**: 789-801

Marmi, J., B. Vila & À. Galobart. 2009. *Solemys* remains from the Maastrichtian of Pyrenees. *Cretaceous Research* **30**: 1307-1312

Matzke, A.T., M.W. Maisch, S. Ge, H.-U. Pfretzschner & H. Stöhr. 2004. A new Xinjiangchelyid turtle from the Qigu Formation. *Palaeontol.* **47**: 1267-1299

Parham, J.F. & J.H. Hutchison. 2003. A new eucryptodiran turtle from the Late Cretaceous of North America. *Journ. Vert. Palaeont.* **23**(4):783-798

Shaffer H.B., P. Meylan & M.L. McKnight. 1997. Tests of turtle phylogeny: Molecular, morphological, and paleontological approaches. *Syst. Biol.* **46**: 235-268

Thomson, R.C. & H.B. Shaffer. 2010. Sparse supermatrices for phylogenetic inference: taxonomy, alignment, phylogeny of living turtles. *Syst. Biol.* **59**: 42-58

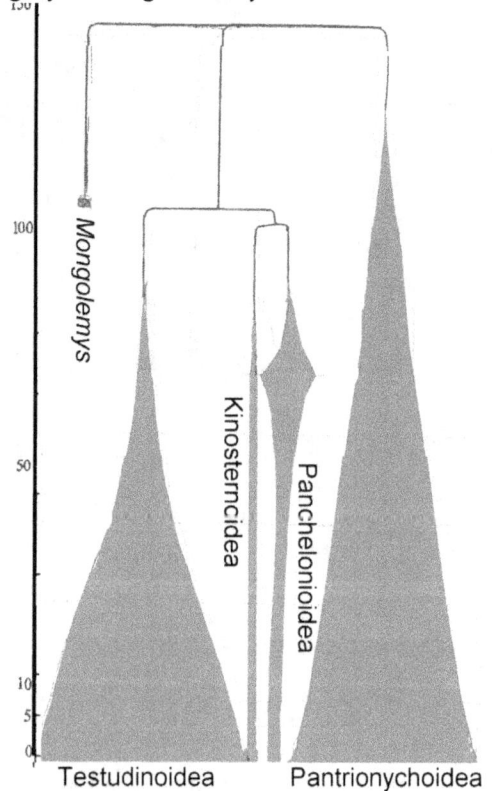

Phylogeny of the living cryptodires.

7. Soft-shelled turtles

Trionychidae are highly distinctive due to their skin-covered carapace. This gives rise to the apt common name of soft-shelled turtles. They all have a flattened shell, with the loss of the peripheral bones. In most species the plastron is thicker than in most other turtles, providing support and protection in their bottom-dwelling habits. They usually lie partially buried in sand or mud, lurking in wait for their prey. The skin covering the body and shell has a rich blood supply, allowing them to take in oxygen from the water and so remain immobile underwater for much longer than most other turtles. They are all largely carnivorous and have a highly specialised skulls adapted for rapid movement and a sharp bite. Some large Asian species of *Pelochelys* and *Chitra* are found in coastal areas and may even venture into the sea.

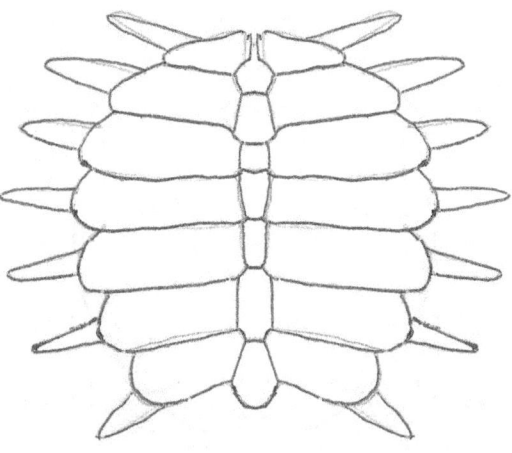

soft-shell turtle carapace bones

The families Dermatemydidae and Kinosternidae are sometimes grouped with the trionychoids and some early trionychoids have been suggested to be close to *Dermatemys* (particularly the genera *Adocus, Basilemys, Nanhsiungchelys* and *Peltochelys*). However, most studies consider the Trionychoidea to form a separate grouping from the other living cryptodires. Despite their relatively weak shell they are one of the most common fossil groups. The most primitive known trionichoid is *Basilochelys macrobios* from Thailand 140 million years ago. This is a relatively complete fossil with parts of a skull, shell and skeleton. In some respects it is a primitive animal but does possess the advanced characters of shell bones with distinctive patterns of ridges on the upper surface, a deep groove in the skull for the internal carotid artery, reduced vertebral scutes, and a wide plastron with wide entoplastral bones all fused to the carapace.

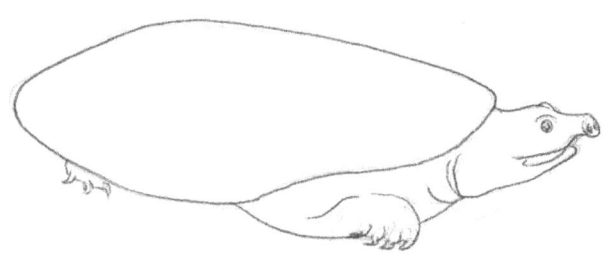

Cyclanorbis
a representative soft-shelled turtle

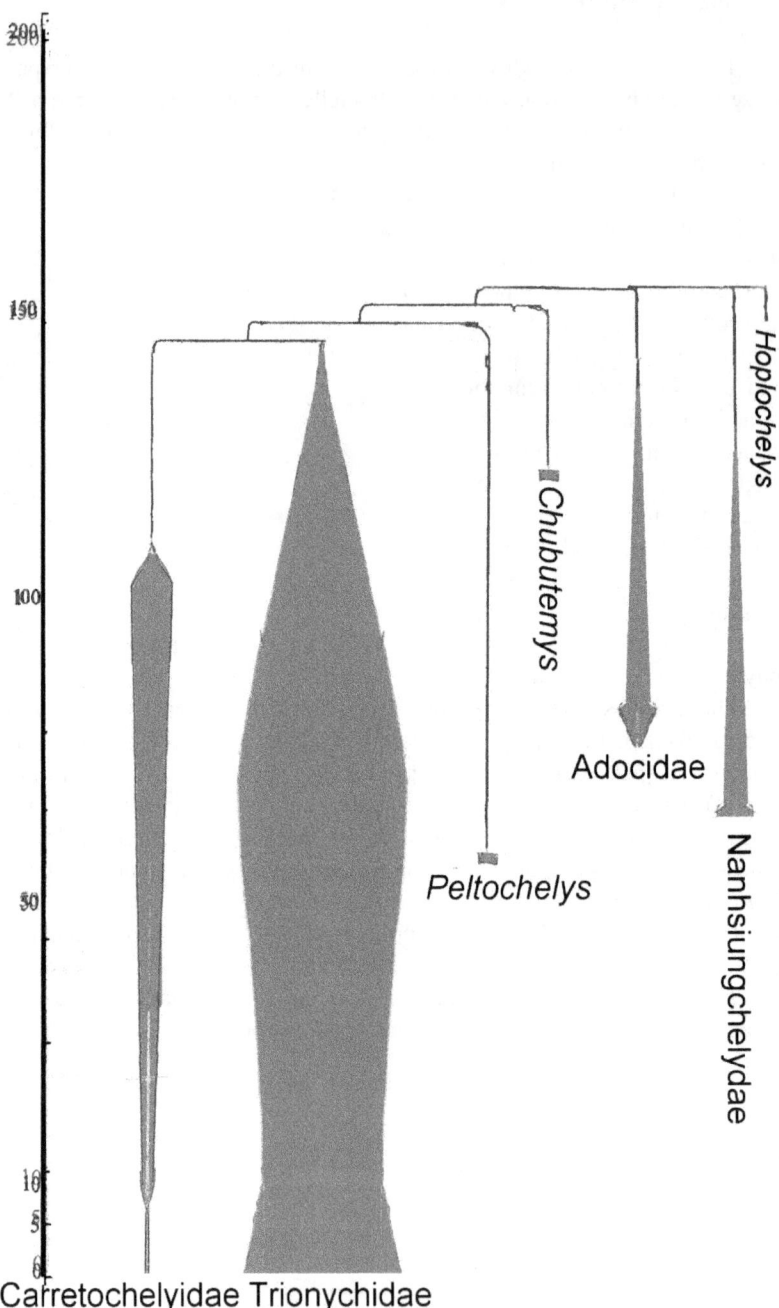

Phylogeny of trionychids.

The soft-shell superfamily Trionichoidea includes the extinct early genus *Hoplochelys* and the extinct families Adocidae and Nanhsiungchelyidae as well as the living Carettochelyidae and Trionychidae. The ancestor of these families appear to have been widespread across much of Laurasia and Gondwana some 160 million years ago, with the Nanhsiungchelyidae and Adocidae originating in Asia. The Adocidae is a moderately diverse family, all of which are extinct. The root of the Adocidae was *Yehguia tatsuensis* which lasted from 155 to 148 million years ago. This was followed by *Adocus* from 100 million years ago to 65 million years ago and *Ferganemys* from 90 million years ago. The small, thin-shelled *Ferganemys* species were replaced by larger, thicker-shelled *Shachemys* due to the 'Turonian transgression' when flooding killed off much of the fauna inhabiting the shallow water and low-lying land which covered much of central Asia, and the appearance of large turtle eating crocodiles and the large predatory ichthyodectid 'bulldog' fish in China. Mongolia was ecologically more stable than middle Asia and there *Shachemys* lasted from 120 million years ago to 80 million years ago.

The Nanhsiungchelyidae seem to have been more terrestrial than the adocids or any living soft-shelled turtles. These were probably species of shallow waters and marshy habitats. They were known from Asia (from Uzbekistan to China) from 130 million years ago, with early species such as *Kharakhutulia kalandadzei* being known from Mongolia. This was followed by three groups: a group of the Asian genera *Hanbogdemys*, *Nanhsiungchelys* and *Anomalochelys*; the Asian *Zangerlia* species; and a group that moved into North America over the shallow seas between Europe and North America 86 million years ago, giving rise to the genus *Basilemys*. This family disappeared 65 million years ago in the mass extinction event that killed off the dinosaurs. The cause of this is still unclear but the general consensus is that an asteroid impact in what is now the Caribbean caused dramatic global environmental change, particularly in low-lying areas, with a mixture of flooding of low-lying terrestrial areas

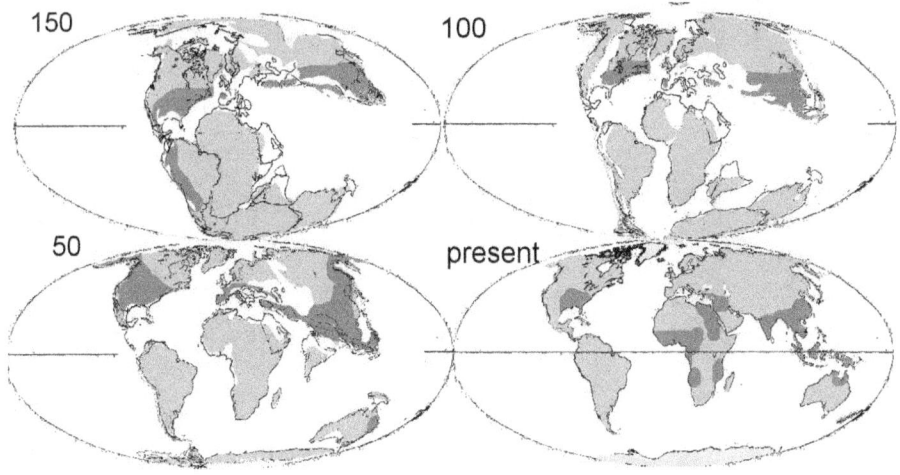

Distribution of trionychoids from 150 million years ago to the present day.

and ecological change in shallow waters. If the Nanhsiungchelyidae were restricted to these environments they would have been particularly badly affected.

Nanhsiungchelyidae were followed by a group containing the rather primitive looking *Chubutemys* and the more advanced trionychids. *Chubutemys copelloi* from Argentina 125 million years ago looks primitive in having a solid skull with little emargination, in comparison to the high degree of emargination in all other soft-shell turtles (both on the skull roof and in the cheek region). The more advanced genera were *Peltochelys* and the living Carettochelyidae and Trionychidae.

The Carettochelyidae comprise one living species and several extinct genera. The only living carretochelyid is the Australian pig-nosed turtle *Carretochelys insculpta*. The family seems to have originated in Laurasia where it was widely distributed and diverse; its arrival in Australia is probably a much more recent event. There were two subfamilies of Carettochelyidae: the Anosteirinae and Carettochelyinae. Anosteirinae comprised the North American *Anosteira* and *Pseudoanosteira*, and the central Asia (Uzbekistan and Mongolia) *Kizylkumemys* from 112 million years ago (with similar forms in Thailand). Carettochelyinae comprised the fossil *Chorlakkichelys* and *Allaeochelys*, and the living *Carettochelys*. *Chorlakkichelys* is represented by a single isolated species from Asia but *Allaeochelys* is a well known genus from North America, Asia and Europe from 54 to 38 million years ago, with many species. The best specimens include *A. crassesculptata* from the Messel Shale of Germany which were deposited in remarkably well preserved 'oilshales' of a 1.5km^2 lake within a volcanic crater. This seems to have been a fresh-water species, but other *Allaeochelys* may have been marine and found throughout the fringes of the Tethys Ocean. Carettochelyinae appear to have survived in Europe until 15 million years ago (specimens from Germany) but were never common in that area.

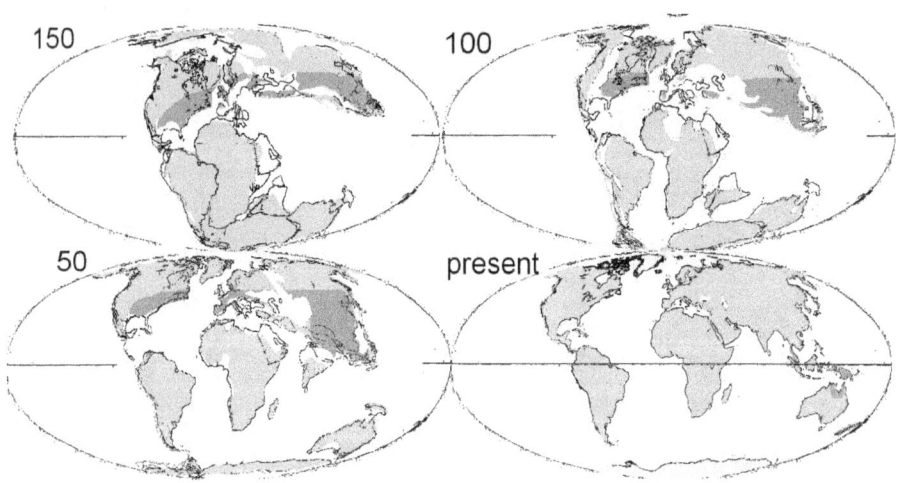

Carratochelyidae distribution from 150 million years ago.

As with most carettochelyids, the ancestor of the pig-nosed turtle *Carettochelys insculpta* probably originated in Asia, moving into New Guinea after the Asiatic and Australian plates collided around 15 million years ago. There is a New Guinean fossil referred to this genus from about 10 million years ago. From 118,000 years ago until about 6,000 years ago what is now the Torres Strait between New Guinea and Australia was a land-bridge, after which time it was flooded by high sea levels. During this time freshwater turtles were able to move between the two areas with the Fly River of New Guinea flowing into Lake Carpentaria (now the Gulf of Carpentaria) and several rivers in Australia. This would have given rise to its present distribution of the southern half of New Guinea and a small area of northern Australia. A Western Australian fossil is probably misidentified. Today it is the largest Australian turtle, reaching 70 cm in length. It is well adapted for fast swimming, with large flippers resembling those of a marine turtle. It does sometimes move into coastal waters but is usually found in rivers and lagoons. It has a very distinctive fleshy snout, which gives it its common name of 'pig-nosed turtle'. This is used to grub in mud to find snails, crustaceans and sometimes fish, although adults feed more on plant matter (especially fruits). The eggs are laid in the dry season, developing normally to start with and then becoming dormant until rising waters stimulate hatching. Hatching may actually occur under water.

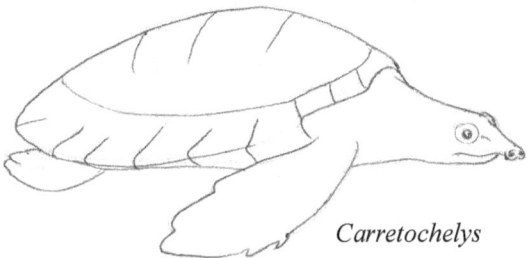
Carretochelys

The Trionychidae were originally North American and Asian and date from 85 million years ago. The earliest fossils include several species of *Aspideretoides* or *Aspiderites*, and *Apalone latus* from about 80 million years ago. The family is divided into three subfamilies: Cyclanorbinae, Trionychinae and the extinct subfamily Plastomeninae. In general the Cyclanorbinae have skin flaps covering the limb openings, the Trionychinae have much less bone in their shells, with much reduced 8th costal bones and plastra. As a purely fossil group the Plastomeninae are much less well defined.

Plastomeninae are known from 84 to 40 million years ago in North America. Four genera are known, *Derrisemys, Hutchemys, Plastomenoides* and *Plastomenus*. These may be ancestral to the Cyclanorbinae, following the movement of a plastomenine across the Bering Land Bridge from North America into Asia around 50 million years ago. The cylanorbinines spread across Asia and into Africa. Today the family contains *Cyclanorbis* and the closely related genera *Cycloderma* and *Lissemys*.

Cyclanorbis is an African genus, with only two species, both found mainly in west Africa: the Nubian flapshelled turtle *C. elegans* and the Senegal flapshelled turtle *C. sengalensis*. The Nubian flap-shell has a fragmented range which is overlapped by the much more widespread Senegal flap-shell. There is some degree of habitat separation; the Senegal flap-shell being found in swamps and stagnant water bodies, with the

Cyclanorbis

Nubian flap-shell mainly in slow moving waters. Both are opportunistic feeders. *Cycloderma* is also African, but with a more southerly distribution: Aubry's flap-shelled turtle *C. aubryi* in central Africa and the Zambezi flap-shell *C. frenatum* from southern and eastern Africa. Both are more active animals than *Cyclanorbis*, living in a wider range of water bodies and being more carnivorous. In contrast to these African genera, the related genus *Lissemys* is Asian, with the Burmese flap-shell *L. scutata* in Burma (and possibly

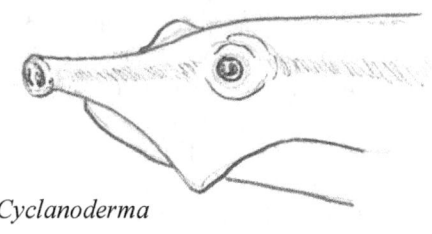

Cyclanoderma

into Thailand and southern China), the Indian flap-shell *L. punctata* found in India, Nepal, Burma and Bangladesh, and the Sri Lankan flap-shell *L. ceylonensis*. The Indian flap-shell has a south-western subspecies *L. p. punctata* in southern India, *L. p. vittata* in the north of India, and the north-eastern *L. p. andersoni* in the Indus, Ganges and Brahmaputra rivers of north India, Bangladesh, Nepal and Burma. These Asian flap--shells are all very similar (although *andersoni* is distinctly spotted) and represent geographical forms that have evolved in isolation in different river systems, with the subspecies forming around 4 million years ago. The separation between the Indian and Sri Lankan flap-shells was earlier, as would be expected from the geography, at about 10 million years ago.

Lissemys

Trionychinae include the exceptional giant fossil species *Drazinderetes tethyensis* from about 45 million years ago in Pakistan (with a shell 80 cm long). Living species are divided into two groups: the large estuarine species (the giant 'Gigantaestuarochelys' – *Pelochelys*, *Chitra* and *Trionyx*) and the others. The other species contain the Apalonina (*Apalone* and *Rafetus*) and the small Amydona (*Nilssonia*, *Aspideretes*, *Amyda*, *Dogania*, *Palea* and *Pelodiscus*).

Although 'Gigantaestuarochelys' is accurately named, translating exactly as 'giant estuarine turtle', it has to be one of the most awkward animal names ever invented. These animals are often more than a metre in carapace length and some species are often found in estuaries, or even out to sea. The giant south-east Asian *Chitra* is found from Pakistan to south-east Asia; fossils from Nepal and India date from about 50 million years ago. Today it comprises three species: the Indian narrow-headed softshell *C. indica* (Pakistan to Bangladesh), Asian narrow-headed softshell *C. chitra* (Thailand) and the critically endangered Burmese narrow-headed softshell *C. vandjiki* (Burma and Thailand). These form a clear geographical pattern, originating in India

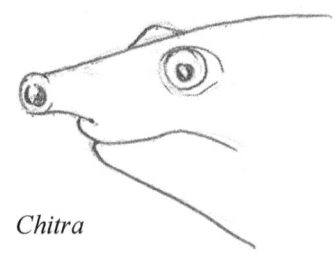

Chitra

(fossil forms and the living Indian narrow-head) and colonising eastwards (Asian and Burmese narrow-heads being very closely related). The Asian narrow-head is divided into the widespread Siamese narrow-headed softshell (*chitra*) and the Javanese narrow-headed softshell (*javanensis*). *Chitra* species are highly aquatic and are found in large water bodies. They are omnivorous, including a large component of animal prey in their diet. They bury themselves in the mud with just their eyes and nostrils exposed and suck in small prey or lunge at larger items. They can be extremely aggressive and the Indian species has been recorded attacking prey as large as goats.

Pelochelys comprises three species, of these the Asian giant softshell *P. cantorii* is distinctive and spread throughout coastal south-east Asia. In New Guinea this species seems to have given rise to two other forms, the northern New Guinea softshell *P. signifera* and the New Guinea giant softshell *P. bibroni*. The two New Guinea species are isolated by the main mountain range that divided New Guinea into distinct north and

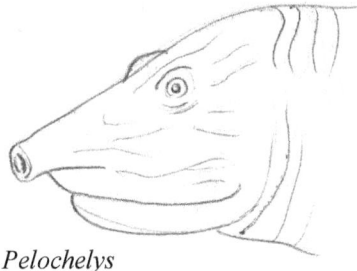

Pelochelys

south areas. The New Guinea giant may reach 1 m in length and the Asian giant 1.3 m and 200 kg; they are heavily built, fleshy animals, often with large rolls of fat.

Trionyx contains only one living species, although there are many fossils. At least 42 *Trionyx* species have been named, but almost all of these belong to other genera. The living species is the Nile soft-shell *T. triunguis* of Africa, and of the fossils probably only the Asian and African ones are true *Trionyx*. Fossils referable to *T. triunguis* are known from 34 million years ago to the present. At about 20 million years ago this species appears to have been widespread around the edges of the Paratethys Sea, a shallow marine area that covered much of eastern Europe and at this time joined the Mediterranean. Around 6 million years ago falling sea levels led to the isolation of the Paratethys and the Mediterranean. These dried out in the

Trionyx

Messinian Salinity Crisis only to be flooded in the 'Zanclean Flood' when the Atlantic Ocean broke into the Mediterranean. The Paratethys remained dry except for its isolated remnants, the Black, Caspian and Aral seas. The drying of the Paratethys and Mediterranean eliminated the European populations of *Trionyx* which was restricted to Africa. The species has been seen at sea and appears to be a good colonist of coastal areas, it has re-colonised the east and north of the Mediterranean relatively recently. This large species (1.2 m) is very adaptable, being omnivorous and found in almost any water body that is associated with nesting sand banks. Nests may contain up to 100 eggs; this high reproductive rate and adaptability mean that it has been able to survive the dramatic changes of the flooding of the Mediterranean and to start to re-colonise its northern shores. Despite this it is now under threat in much of its range due to high levels of hunting in recent years.

Apalonina contains two very different genera, the relatively small North American soft-shells (*Apalone*) and the gigantic Asian *Rafetus*. This relationship is rather questionable, being found in some studies but not others (e.g. not supported by mitochondrial DNA); in morphology the two genera share the reduction or loss of the 8th pleural bones. *Apalone* is now exclusively North American, but there is at least one fossil species from South America. The evolution of the *Apalone* species is unclear, with different studies giving contradictory results. Of the living species, some studies consider the smooth softshell *A. mutica* of much of the USA and the spiny softshell *A.*

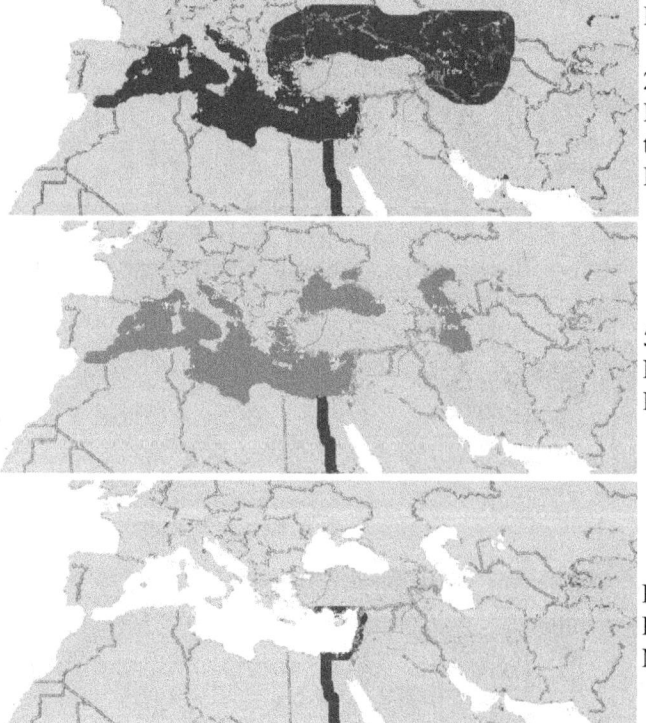

Range change in *Trionyx*.

20 million years ago
Maximum range in the Mediterranean and Paratethys.

5 million years ago
Mediterranean and Paratethys largely dry.

Present day
Recolonising the eastern Mediterranean.

spinifera from Canada to Florida to be more closely related to one than to the Florida softshell *A. ferox* of the Gulf states of the USA, but others consider the spiny and Florida softshells to be sister species. This latter arrangement has slightly more support. Mitochondrial DNA studies show that both smooth and spiny softshells show genetic differences between north-west and

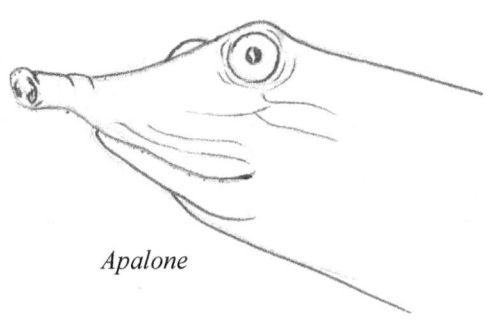

Apalone

south-eastern populations, whereas the Florida softshell is genetically more uniform. From 5 to 3.5 million years ago Florida would have been an island surrounded by higher sea-levels and other refugia would have been created in different parts of North America. *Apalone* species were probably isolated in these refugia, the ancestor of the Florida softshell on the Florida island and smooth and spiny softshell on more than one refugium. Expansion of ranges after sea-levels had fallen would have resulted in a more widespread, adaptable Florida softshell and the presence of different genetic races of the other species. The Florida softshell is highly adaptable and tolerant of extreme variations in water salinity and temperature. Its diet is varied although some old animals specialise in feeding on snails and develop broad, crushing jaws as a result. The smooth softshell seems to be related to the 80 million years ago *A. latus* from further north in North America and is divided into the midland smooth softshell *A. m. mutica* and the Gulf Coast smooth softshell *A. m. calvata*. The spiny softshell has several subspecies in different rivers. Before the Ice Ages the rivers running through the central and northern parts of the Great Plains of North America were isolated from those of the western Gulf. The wide-ranging spiny softshell would have diverged into two forms; a north-eastern form in the Great Plains rivers and a western one around the Gulf of Mexico. During the Ice Ages and the periods between them river courses changed, allowing both mixing of populations and isolation at different times. The different river populations would have diverged genetically in isolation. This would have given rise to the northern and eastern subspecies: eastern *A. s. spinifera*, Gulf coast *A. s. aspera* and western *A. s. hartwegi* and the further western subspecies: black spiny softshell *A. s. atra* of Mexico, the Texas spiny softshell *A. s. emoryi* of Texas and Mexico, Guadalupe spiny softshell *A. s. guadalupensis* of Texas and pallid spiny softshell *A. s. pallida* of Louisiana to Texas. As climatic conditions changed these populations would have come into close contact with other forms as river flow altered. For the northern populations in particular a major change would have been the rivers that used to drain into the Hudson Bay becoming re-routed into the Mississippi river. As a result the most northerly subspecies *A. s. spinifera* and *A. s. hartwegi* have become isolated by at least 2500 km but are genetically extremely close, whereas *A. s. hartwegi* and *A. s. pallida* are much more distinctive genetically but may be less than 200 km apart. In the Gulf region the Red and Pecos Rivers remained isolated from the Cimarron and Arkansas rivers; this isolation resulted in a division between southern and northern subspecies. The southern pair are the Rio Grande – Pecos River subspecies *A. s. emoryi* and *atra* (Cuatro Ciénegas). The nothern pair are mainly from the Red River: *A. s. pallida* (Oklahoma, to Texas) and *A. s. guadalupensis* (restricted to Texas where it is found in the Red River and other rivers).

Rafetus contains the Euphrates soft-shell *R. euphraticus* and the ugliest of all living turtles, the Shanghai soft-shell *R. swinhoei*. This is also known as the Red River, Yangtze or Swinhoe's giant soft-shell. It is the largest living fresh-water turtle, with a shell of over 50 cm, a body length of over a metre and weighing up to 150 kg. The Vietnamese population was considered a distinct species at one time, but the Hoan Kiem turtle *R. leloii* is identical to

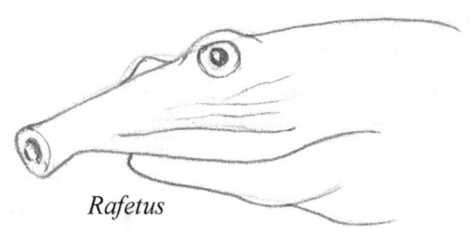

Rafetus

the Chinese animals. The species has been associated with Hoan Kiem Lake in Hanoi for hundreds of years. The first recorded sighting was in the reign of King An Duong Vuong (257-207BC) when the Golden Turtle God Kim Qi emerged from a lake to defend the ancient capital of Vietnam from invasion by directing the building of the defences and making a magical cross-bow that fired sheaves of arrows at every shot. Once the kingdom was saved from invasion Kim Qi returned to the lake, emerging again to warn the king of a conspiracy involving his daughter. She was executed and the king drowned himself in the lake. Kim Qi's later appearance was less tragic. In the 15th century a fisherman pulled a magical sword out of the lake, this was given to a peasant called Lê Lợi who used it to lead a rebellion against the Chinese occupiers of Vietnam. After becoming the first Lê emperor Lê Lợi threw the sword back into the lake, where it was caught by Kim Qui. From then on the lake has been called Hoan Kiem, 'Lake of Returning Sword'. Hoan Kiem Lake is very shallow (mostly less than a metre) and is now very heavily polluted; only one giant soft-shell is thought to survive there. It is very rarely observed. In 2011 it was seen to have developed a number of infected lesions on different parts of the body and was captured for treatment before being returned to the lake.

The Shanghai soft-shell originally inhabited the Yangtze River and Lake Taihu in eastern China, Gejiu, Yuanyang, Jianhshui and Honghe in Yunnan province of southern China, the Red River of northern Vietnam. However, it seems to have disappeared from most of this range. The last sighting in China was in 1998 when one was found in the Red River. By 2004 only a few captive survivors were known; a male in Beijing Zoo died in 2005, as did another male at Shanghai Zoo in 2006 (both of these came from Gejiu in the1970s). This left only two surviving captive animals in China: an 80-year old female in Changsha Zoo and a 100-year-old male in Suzhou Zoo. The female was moved to Suzhou in 2008 in an attempt to breed them. She laid eggs every year from 2008, but so far almost all have been infertile and the few fertile eggs have failed to hatch. Vietnam still has two surviving animals (both males) - one in Hoan Kiem Lake and one in Dong Mo Lake outside Hanoi. In 2008 the dam containing the Dong Mo Lake burst and the turtle was washed into the Red River. This animal was recaptured after a few days and returned to the lake. There are plans to dam the Red River in China which could be catastrophic for any undetected survivors.

    Amydona species are all small to medium sized Asian species, with a distinctive ridge at the symphysis of the lower jaw. The earliest genus in the group is *Pelodiscus*; this is followed by *Palea, Dogania, Amyda* and finally the moderately diverse grouping of *Apsideretes* and *Nilssonia*. *Pelodiscus* is largely Chinese and has an unknown number of species. Most *Pelodiscus* have been put in the Chinese softshell *P. sinensis*, but four species may be valid. A genetic study of a small number of individuals, some of uncertain origins, supported the view that several species are present in the genus. Of the possible forms the Northern Chinese softshell *P. mackii* is the most distinctive, with a pattern of pale mottling on the skin and reaching up to 33 cm, compared to the 20 cm of other *Pelodiscus*. These other less distinctive forms are the Hunan softshell turtle *P. axenaria* of Hunan province, China, the lesser

Amydona

Chinese softshell *P. parviformis* (Guangxi province, China) and the Chinese softshell *P. sinensis* (throughout China and south-east Asia, Japan and Taiwan, also widely introduced). The lesser Chinese softshell is the smallest, at only 10-12 cm.

    The Chinese softshell has the shortest incubation duration known for any turtle – just 28 days, although it is usually 40-80 days. This used to be a highly successful species with a wide climatic tolerance, varied habitat preferences and rapid reproduction (being capable of reaching maturity within one year). However, it is now consumed for food in very large numbers in China. Millions of turtles are produced each year in farms but the species seems to have disappeared from much of its wild range.

    Palea is restricted to the wattle-necked softshell *P. steindachneri* of China, Laos and Vietnam. It is an adaptable species found in a wide range of rivers and lakes, often at high altitude. This adaptability has resulted in successfully established introduced populations on the islands of Mauritius and Hawaii.

    Amyda contains the Asiatic softshell *A. cartilaginea* of marshlands in India to Vietnam, Borneo and Sumatra. This is similar to the wattle-necked softshell but more specialised in its habitat preferences and hence

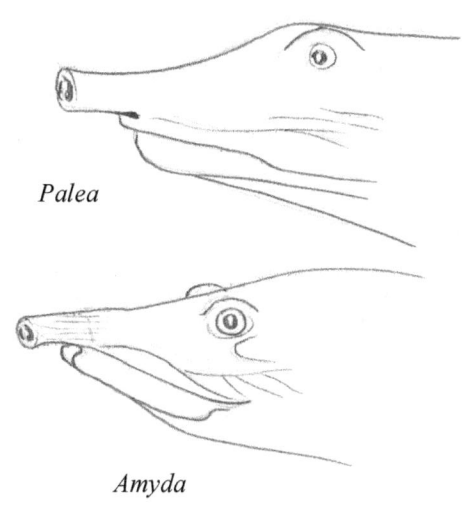

Palea

Amyda

less adaptable.

Dogania is restricted to the forest dwelling Malayan softshell turtle *D. subplana* of Borneo, Sumatra, Java, Philippines, Malay peninsula.

Of the *Aspideretes/ Nilssonia* species the Indian softshell *A. gangeticus* (Afghanistan to Bangladesh) is the most isolated species, followed by the Indian peacock softshell *A. hurum* (Pakistan to Bangladesh) and three closely related species: Leith's softshell *A. leithi* (India), black softshell *A. nigricans* (east India and Bangladesh) and the Burmese peacock softshell *Nilssonia formosa*.

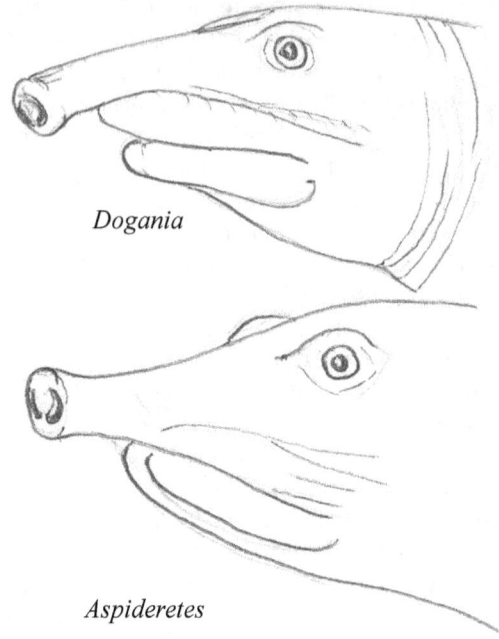

Dogania

Aspideretes

In addition to these there are many extinct early trionychine forms, including some species placed in "*Amyda*" and "*Trionyx*" as well as the extinct genera *Platypeltis* and *Plastomenus*; most of these are known from the past 5 million years and two living genera include fossil forms (two species of *Chitra*, at least six of *Cyclanorbis*). Some were relatively small, such as *Axestemys* from 60 million years ago most of which were 45 cm long. This genus was primarily North American and was moderately diverse. One species from France, *A. vittatus*, has been attributed to the genus. Fossils of other members of the group are known from Asia, Europe, Africa, North America, South America and Australia. The living species persist in all these areas except Australia. Fragmentary fossils are known from Queensland from about 40 million years ago, disappearing from the continent some 40,000 years ago. Only one is reliably identifiable: *Murgonemys braithwaitei* is known from a nearly complete carapace and a partial plastron, from somewhere between 56 and 34 million years ago in southeastern Queensland. This can be identified as a member of the Trionychinae due to its typical trionychoid shell (lack of scutes, sculptured shell bones, lack of fusion between carapace and plastron and lack of suprapygal and peripheral bones) and specifically as a trionychid based on the form of the bones in the plastron. However, it has an expanded nuchal with a unique shape and seven neural bones in the carapace, which are not normal trionychid features. *Murgonemys* is the oldest southern hemisphere trionychid, and the family survived on the continent until about 40,000 years ago. Trionychids must have reached Australia from Asia despite an enormous ocean crossing at the time. A South American origin is even more unlikely. Living large soft-shells (*Chitra* and *Pelochelys*) sometimes venture into the ocean and the ancestor of *Murgonemys* probably did the same, enabling it to cross open ocean to Australia.

References

Danilov, I.G. 1999. The ecological types of turtles in the Late Cretaceous of Asia, *Proc. Zool. Inst. RAS* **281**: 107–112

Engstroma, T.N., H.B. Shaffer & W.P. McCord. 2002. Phylogenetic diversity of endangered and critically endangered southeast Asian softshell turtles (Trionychidae: *Chitra*). *Biol. Conserv.* **104**: 173-179

Engstrom, T.N., H.B. Shaffer & W.P. McCord. 2004. Multiple Data Sets, High Homoplasy, and the Phylogeny of Softshell Turtles (Testudines: Trionychidae). *Syst. Biol.* **53**(5):693–710

Fritz, U., S. Gong, M. Auer, G. Kuchling, N. Schneeweiß & A.K. Hundsdörfer. 2010. The world's economically most important chelonians represent a diverse species complex (Testudines: Trionychidae: *Pelodiscus*). *Org. Divers. Evol.* **10**

Gardner, J.D., A.P. Russell & D.B. Brinkman. 1995. Systematics and taxonomy of soft-shelled turtles (Family Trionychidae) from the Judith River Group (mid-Campanian) of North America. *Can. J. Earth Sci.* **32**: 631–643.

Güçlü, O., C. Ulger, O. Türkozan, R. Gemel, M. Reimann, Y. Levy, S. Ergene, A.H. Uçar & C. Aymak. 2009. First Assessment of Mitochondrial DNA Diversity in the Endangered Nile Softshell Turtle, *Trionyx triunguis*, in the Mediterranean. *Chel. Conser. Biol.* **8**(2): 222–226

Hirayama, R., D.B. Brinkman & I.G. Danilov. 2000. Distribution and biogeography of non-marine Cretaceous turtles. *Russ. Journ. Herpetol.* **7**(3): 181198

Iverson, J.B., R.M. Brown, T.S. Akre, T.J. Near, M. Le, R.C. Thomson & D.E. Starkey. 2007. In Search of the Tree of Life for Turtles. *Chel. Res. Monogr.* **4**: 85–106

Joyce, W.G., N. Klein & T. Mörs. 2004. Carettochelyine Turtle from the Neogene of Europe. *Copeia* **2004**(2): 406-411

Joyce, W.G. & T.R. Lyson. 2010. A neglected lineage of North American turtles fills a major gap in the fossil record. *Palaeontology* **53**: 241–248

Karl, H.-V. 1999. Die Zoogeographie der känozoischen Weichschildkröte *Trionyx triunguis* Forskal, 1775 (Testudines: Trionychidae). *Joannea Geol. Paläont.* **1**: 27–60

McGaugh, S.E., C.M. Eckerman & F.J. Janzen. 2008. Molecular phylogeography of *Apalone spinifera* (Reptilia, Trionychidae). *Zool. Script.* **37**: 289–304.

Meylan, P.A. 1987. The phylogenetic relationships of soft-shelled turtles (family Trionychidae). *Bull. AMNH* **186**: 1–101

Mlynarski, M. 1976. *Handbuch der Paläoherpetologie. Band 7. Testudines.* Gustav Fischer Verlag, Stuttgart & New York

Nessov, A. 1977. A new genus of pitted-shelled turtle from the upper Cretaceous of Karakalpakia. *Palaeont. J.* **1**: 96-107

Praschag, P. H. Stuckas, M. Packert, J. Maran & U. Fritz. 2011. Mitochondrial DNA sequences suggest a revised taxonomy of Asian flapshell turtles (*Lissemys* Smith, 1931) and the validity of previously unrecognized taxa (Testudines: Trionychidae). *Vertebr. Zool.* **61**(1): 147–160

Romer, A.S. 1956. *Osteology of the Reptiles.* University of Chicago Press

Shaffer H.B., P. Meylan & M.L. McKnight. 1997. Tests of turtle phylogeny: Molecular, morphological, and paleontological approaches. *Syst. Biol.* **46**: 235-268

Sukhanov, V.B., I.G. Danilov & E.V. Syromyatnikova. 2008. The description and phylogenetic position of a new nanhsiungchelyid turtle from the Late Cretaceous of Mongolia. *Acta Palaeontol. Pol.* **53**(4): 601–614.

Syromyatnikova, E.V. 2011. Turtle of the genus *Ferganemys* Nessov et Khosatzky, 1977 (Adocidae): shell morphology and phylogenetic position. *Proc. Zool. Inst. RAS* **315**: 38–52

Thomson, R.C. & H.B. Shaffer. 2010. Sparse Supermatrices for Phylogenetic Inference: Taxonomy, Alignment, Rogue Taxa, and the Phylogeny of Living Turtles. *Syst. Biol.* **59**(1): 42–58

Tong, H., J. Claude, W. Naksri, V. Suteethorn, E. Buffetaut, S. Khansubha, K. Wongko & P. Yuangdetkla. 2009. *Basilochelys macrobios* n. gen. and n. sp., a large cryptodiran turtle from the Phu Kradung Formation (latest Jurassic-earliest Cretaceous) of the Khorat Plateau, NE Thailand. *Geological Society, London, Special Publications* **315**: 153-173

Vitek, N.S. & I.G. Danilov. 2010. New material and a reassessment of soft-shelled turtles (Trionychidae) from the Late Cretaceous of Middle Asia and Kazakhstan. *Journ. Vert. Paleont.* **30**: 383-393

Walter, J.G. & T.R. Lyson. 2010. A neglected lineage of North American turtles fills a major gap in the fossil record. *Palaeontology* **53**: 241-248

Weisrock, D. & F.J. Janzen. 2000. Comparative Molecular Phylogeography of North American Softshell Turtles (*Apalone*): Implications for Regional and Wide-Scale Historical Evolutionary Forces. *Mol. Phyl. Evol.* **14**: 152–164

White, A.W. 2001. A new Eocene soft-shelled turtle from Murgon, south-eastern Queensland. *Mem. Ass. Austral. Palaeontol.* **25**: 37-44

8. Sea turtles

The sea turtles (Chelonioidea) have a long history, dating from 145 million years ago, although the living families Cheloniidae and Dermochelydiae are comparatively recent. The oldest definite cheloniid is *Procolpochelys* from 15 million years ago in North America, although *Puppigerus* and *Euclastes* and the earlier (65 million years ago) *Toxochelys* and *Ctenochelys* from North America may also belong to this family. Dermochelyd fossils from North America and Japan date from 80 million years ago and include *Eosphargis* and *Psephophorus* from 56 to 34 million years ago of Europe and *Egyptemys* of North Africa and North America 55 million years ago.

Sea turtles show several highly distinctive adaptations: the body is adapted for swimming, especially with the modification of the fore-limbs into paddles and the hind-limbs into rudders, and the jaws of many turtle species have become adapted for specialised feeding. The feeding adaptations are mainly those for 'durophagy' (shearing or crushing). In living turtles shearing is associated with herbivory, particularly feeding on seagrasses in the green turtle *Chelonia mydas*. Crushing is seen in snail crushers such as the loggerhead *Caretta caretta*. In addition there is one specialised jellyfish predator, the leatherback *Dermochelys coriacea*. This shows few clear skeletal adaptations to this diet other than a distinctively hooked and notched upper jaw which would be good for gripping soft, slippery prey. These modifications develop progressively through the fossil record of the sea turtles.

The sea turtles divide into two groups, the leatherback lineage Dermochelyoidea and the advanced sea turtles, the Cheloniidae. All chelonioid families tend to show the same evolutionary trends: increase in size of fontanelles in the carapace and plastron, hind limb modification into 'rudders' (reduction in size) and forelimbs into 'paddles' (humerus straightened and flattened, digits immobilized) and a shift in the jaw so that the mandibular coronoid process (the attachment site of jaw closing muscles) is moved

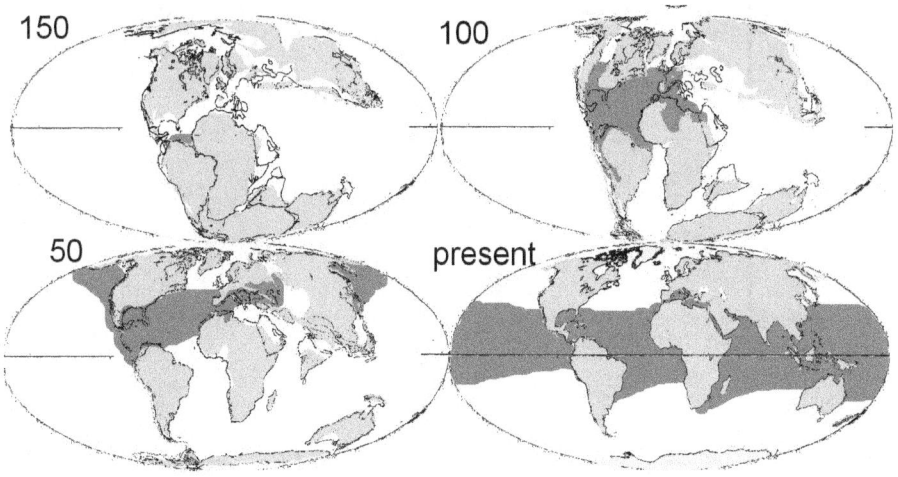

Distribution of sea turtles from 150 million years ago.

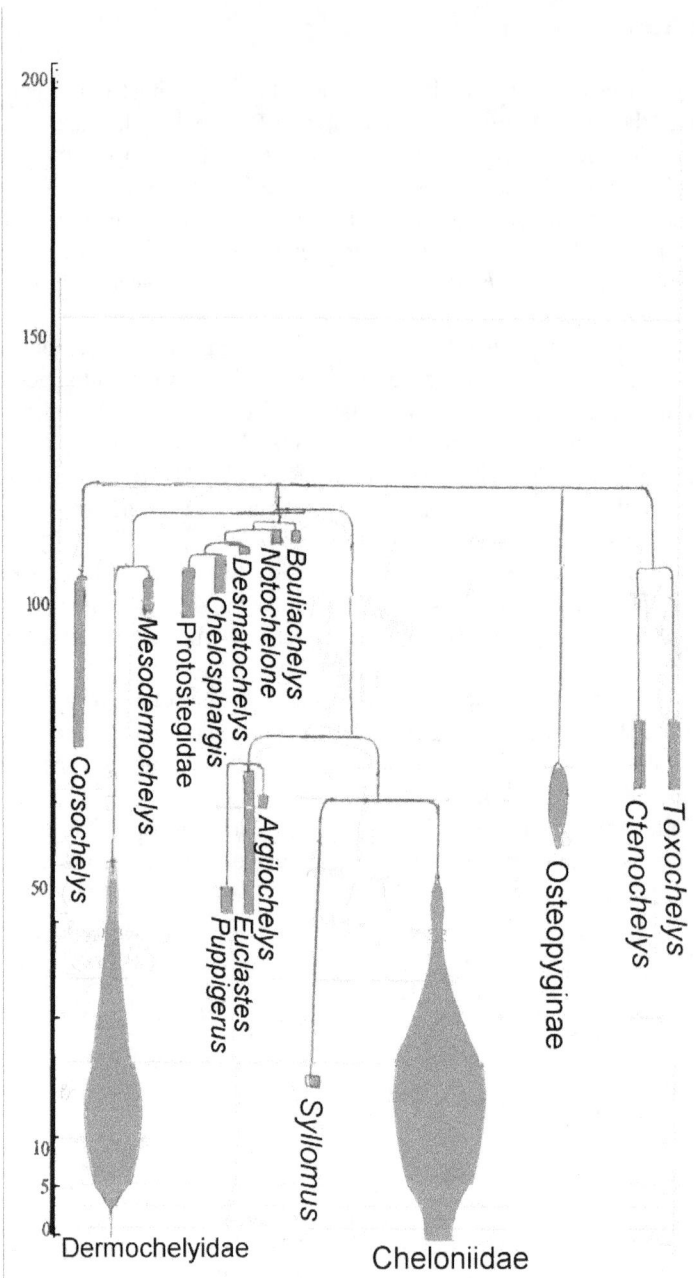

Phylogeny of sea turtles.

backwards, compressing the posterior part of the jaw. These changes are associated with an increasingly pelagic (open water) life. The early species have small plastra, with narrow bridges. These are associated with only slight modification of the limbs.

The dermochelyds have notably reduced shell bones, reaching an extreme in the living leather-backed turtle *Dermochelys coriacea*. The shell contains several large openings, these fontanelles are apparent even in primitive species (*Desmatochelys* and *Mesodermochelys*), which also have reduced contact between the hyo- and hypoplastral elements (reduced to a narrow projection). They also tend to have very long fore-limbs. Dermochelyds contained the Protostegidae and the Dermochelyidae. Protostegidae is an extinct family known from 140 to 65 million years ago. It included *Santanachelys*, *Terlinguachelys* and derived protostegids. Early Dermochelydiae included *Mesodermochelys*.

Many Protostegidae were giant sea turtles, exceptionally large animals with reduced shells. In some cases fossils contain shellfish remains. Some specimens of *Protostega gigas* have tooth marks from large sharks and in one case the shell has teeth of the shark *Cretoxyrhina mantelli* embedded in them. The earliest protostegid is *Santanachelys gaffneyi* from Brazil 140 million years ago. *Santanachelys* retained relatively short forelimbs; as with other protostegids the digits in the flippers were not fused, unlike in living sea turtles, although there was a tendency towards immobilisation in more advanced species. *Terlinguachelys fischbecki* from 80 million years ago in North America retained a number of primitive characters, including a well ossified plastron and costals, with small fontanelles, elongate femora and mobile digits. *Archelon ischyros* from 74 million years ago from North America was a massive animal, reaching up to 4.5 m in total body length and with flippers spanning 5.25 m. The skull was large, with a powerful hooked beak thought to be adapted for eating squid and nautiloids. One specimen is fossilized in a very unusual posture, with its limbs folded and head bowed and is thought to have died while hibernating in the mud of the sea floor.

The distinctively notched hooked upper jaw of several protostegids (such as *Protostega* and *Bouliachelys suteri*) may have been an adaptation to feed on the abundant ammonites of the period. Secondary palates developed in some of these species, but these were always much less developed than in most of the living sea turtles. The secondary palate in these species would have strengthened the jaw when shearing through the hard shells of the ammonites.

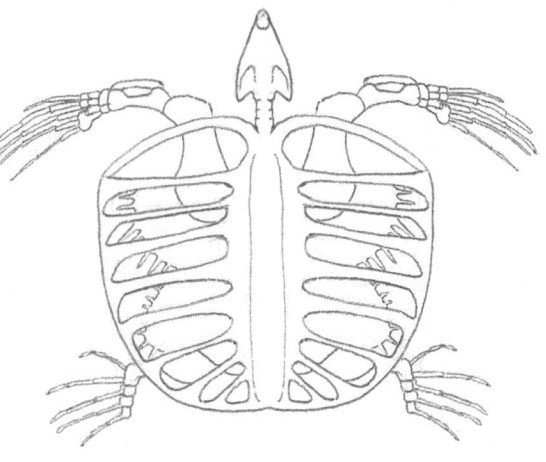

Archelon

The North American protostegid fossils are known from the 'Western Interior Seaway' that divided North America from 145 to 100 million years ago, flowing from north-central Canada to the Gulf of Mexico. Protostegids became extinct in the Cretaceous-Tertiary mass extinction event 65 million years ago when disruption of the shallow marine ecosystems would have caused dramatic changes to the marine environment of what remained of the Western Interior Seaway, including the extinction of many of the marine invertebrates that the protostegids fed upon.

The protostegids were replaced by the dermochelyds which feed largely on jellyfish, which remained abundant. The Dermochelyidae have very long forelimbs, typically much longer than the hindlimbs; this is the case even in the earliest dermochelyid *Mesodermochelys undulatus*. Dermochelyds appear to have decreased in diversity, from at least eight forms 40 million years ago (placed in *Cosmochelys, Egypemys, Eosphargis* and "*Psephophorus*"), four 30 million years ago (in *Natemys* and "*Psephophorus*"), two about 20 million years ago (*Psephophorus*) to only one modern species. This living survivor is the leatherback turtle *Dermochelys coriacea*, a specialist predator of jellyfish and comb-jellies. The beak has a well-developed notched hook, similar to that seen in *Protostega* which allows gripping of the soft bodies of the jellyfish. More notably the throat is lined with long, backward pointing spines which ensure that the slippery food is passed down to the stomach. This is the largest of the living sea turtles, sometimes exceeding 3 meters in carapace length. The leatherback is the only living sea turtle to lack a secondary palate. With its soft food diet it has no need for strong jaws and the bones of the skull are surprisingly loose. In many ways the leatherback turtle is the most remarkable of the living sea turtles, but it is probably fairly typical of a dermochelyd. Its great size means that it retains a stable body temperature even when external temperatures fluctuate. This allows it to have the greatest geographical range of any living turtle, from tropical waters to the cold waters off Greenland. At this northern latitude they are able to retain 20°C body temperatures despite the water being only 7°C. This characteristic would have been

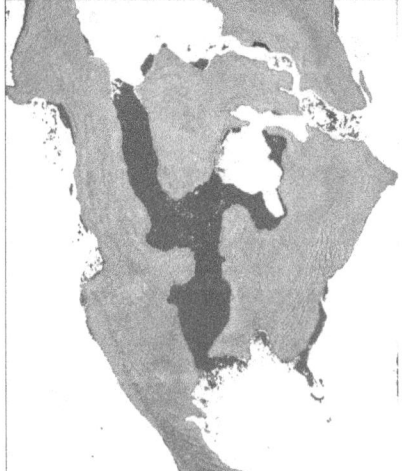

The North American Western Interior Seaway (dark grey) 140 million years ago.

found in almost any large sea turtle, such as most of the dermochelyds and many of the other gigantic fossil forms. The cold tolerance of the leatherback allows it to be active in the southern oceans as well, enabling it to move around the southern tip of Africa and this prevents the Indian Ocean populations

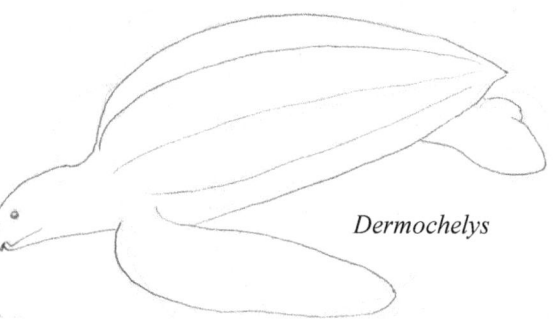

Dermochelys

diverging from the Atlantic ones. Consequently leatherbacks are genetically relatively uniform across their vast range, being the only wide ranging sea turtle with no recognised subspecies.

The earliest Cheloniidae relatives were the extinct Toxochelyidae which lived from 84 to 55 million years ago. These included species of *Toxochelys* (*T. latiremis* and *T. moorevillensis*), *Catapleura repanda* (=*Toxochelys atlantica*), *Dollochelys, Porthochelys laticeps* and *Thinochelys lapisossea*. These were widespread generalists, lacking any of the feeding specialisations for herbivory or durophagy. Toxochelyids were followed by the Lophochelyninae. These were also generalists living from 84 to 65 million years ago but at around 80 million years ago *Mexichelys coahuilaensis* was the first known crusher. The extinction of the protostegids 65 million years ago would have allowed the cheloniid relatives ('pancheloniids') to occupy the now vacant crushing niche, and to diversify rapidly. The origin of herbivory in cheloniids may be linked to the development of seagrass communities, which first appeared around 80 million years ago and by 65 million years ago were widespread. At 70 million years ago shearing had evolved, with the cheloniid *Allopleuron hofmanni* and the dermochelyd *Mesodermochelys undulatus*.

Shearing seems to have evolved repeatedly in the cheloniids as indicated by the shearing *Argillochelys* from 55 million years ago being related to the west Russian generalist *Itilochelys rasstrigin* from 65 million years ago. *Itilochelys* does not have the fully pelagic specialisations of the fore-limbs, seemingly being a coastal species. *Eochelone* from 40 million years ago also remained a generalist.

More recent cheloniids include '*Euclastes*', *Puppigerus* and the true Cheloniidae. '*Euclastes*' has been applied to several species, apparently in two very closely related genera: *Euclastes* and *Pacifichelys* (*P. urbani* of Peru and *P. hutchinsoni* from North America, both from 15 million years ago). These were crushers but show only slight modifications of the limbs. *Euclastes* (including *E. gosseleti*) was a coastal genus, known from 70 million years ago, in the margins of the Tethysian and Atlantic Oceans, disappearing in the mass extinction event 65 million years ago. It had an extensive secondary palate, associated with strengthening the jaws for crushing, probably feeding on shallow water shelled invertebrates which suffered severely in the mass extinction.

True cheloniids evolved the fully developed 'rudders' and 'paddles' typical of living sea turtles; these are all open water species, with the exception of the living flatback turtle *Natator depressus*. They comprise two groups: the Carettini (hawksbill, ridley and loggerhead turtles) and the Cheloninii (green and flatback) which may have separated around 50 million years ago. At this time the continents were broadly in the positions they are in today, but wide sea connections existed between the Atlantic, Pacific and Indian Oceans. The first evidence of the Carettini comes from around 14 million years ago, with fossils from North America. This tribe comprises the highly divergent hawksbill (*Eretmochelys*), ridley (*Lepidochelys*) and loggerhead (*Caretta*). The crushing mode of feeding has probably evolved separately in loggerheads and ridleys. All of the Cheloniidae are closely related and hybrids have been recorded between most species, even between members of the Carettini and Cheloninii (green×hawksbill hybrids from Surinam and Mexico, green×loggerhead from Brazil, Canada and Australia and green×ridley from Brazil). The Carettini species diverged around 10 million years ago. They remain genetically relatively close and hawksbill×loggerhead hybrids have been recorded from Brazil, the USA, Bahamas, Japan, and China, hawksbill×ridleys from Brazil and the USA, and loggerhead×ridley in Brazil and the USA. It is not known if these hybrids are fertile.

Lepidochelys are the ridley turtles, with two species: the olive ridley *L. olivacea* and Kemp's ridley *L. kempii*. The olive ridley is the smallest of the living sea turtles with a shell length of less than 74 cm. It mainly feed on molluscs, crustaceans, jellyfish, sea urchins and fish. Some seagrasses and seaweeds are consumed. This species is found throughout the Indo-Pacific region, mainly tropical but also feeding in non-tropical waters. Olive ridleys have a mass-nesting strategy ('arribadas') when tens of thousands of females may emerge to nest at the same time. These arribadas may serve to swamp potential predators (birds, racoons, foxes etc) and ensure that some hatchlings survive. Arribadas occur in Costa Rica, Mexico and India, but there are also nesting areas where small numbers breed. Why some populations breed in arribadas and others do not is not clear. The Kemp's ridley has a single population restricted to the Atlantic Ocean. This is a rare species breeding in arribadas along the coast of

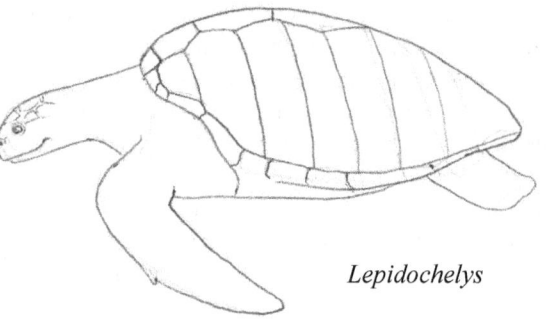

Lepidochelys

the Gulf of Mexico, mainly in Texas and Mexico. Due to over-hunting, accidental capture in fisheries and pollution the numbers are greatly reduced now and the arribadas are now mostly reduced to hundreds of females.

The loggerhead turtle *Caretta caretta* is related to one fossil species, *C. patriciae* from 5-3 million years ago. Both species have similar powerful crushing jaws which are used in the living loggerhead to crush the shells of marine snails. The loggerhead turtle ranges widely, nesting as far north as 40°. These are the largest of the hard-shelled sea turtles, with a shell length of over 2 metres. They are unusually

90

aggressive for sea turtles, with females apparently being territorial. This probably arises from their diet, with feeding areas being held as territories until the local supply of prey has been exhausted. The living species has two subspecies: the Atlantic *C. c. caretta* and the Indo-Pacific *C. c. gigas*. The Atlantic population breeds mainly along the

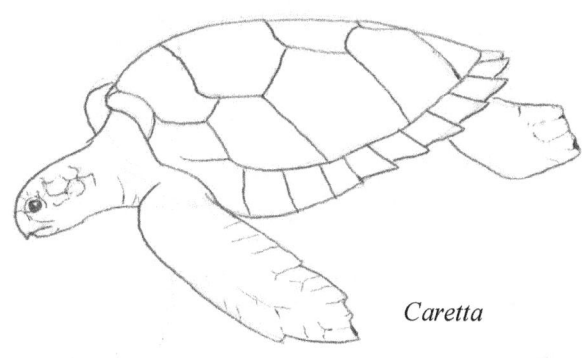

Caretta

coastline of the Gulf of Mexico but some also breed in the Mediterranean and on the African and South American coasts. Indo-Pacific loggerheads breed in scattered colonies in southern Africa, Arabia (Oman), Australia, Japan, and occasionally on Pacific islands and the west coast of North America.

Eretmochelys contains a single species: the hawksbill turtle *E. imbricata*. The hawksbill is the only living species of turtle to feed largely on sponges. The pointed, hooked beak that gives the species its name is well adapted to cutting sponges and other immobile, soft-bodied animals (sea anemones and sea squirts) off rocks and reefs. The hawksbill is a critically endangered species, having been under very heavy hunting pressure for its shell. The scutes are beautifully patterned and have been exploited for 'tortoise shell'. The trade in this is now restricted and illegal in most countries. There is still some poaching but the pressures on this species are much reduced.

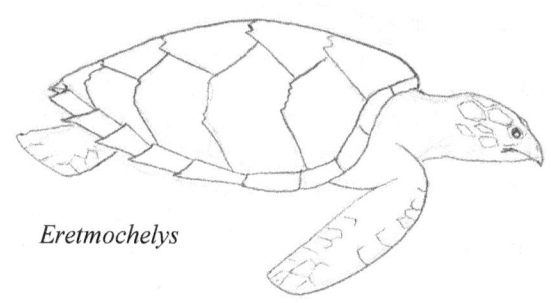

Eretmochelys

They are now most threatened by development on their nesting beaches. The hawksbill has two subspecies: *E. i. imbricata*, of the Atlantic Ocean and *E. i. bissa* of the Indian and Pacific Oceans. Hawksbills are almost entirely tropical and rarely nest outside of equatorial latitudes.

In addition to the living species two extinct species are placed in the Cheloninii: *Procolpochelys grandaeva* and *Syllomus aegypticus* from around 18 million years ago and 8 million years ago respectively.

Natator has often been grouped with the Carettini but molecular data strongly support its grouping with *Chelonia*. *Natator* is apparently related to *Syllomus*, both having well-ossified plastra. *Syllomus* had shearing jaws and had converged on the same herbivorous life as *Chelonia*. Fossils of *Chelonia* are known from 5 million years ago to the present.

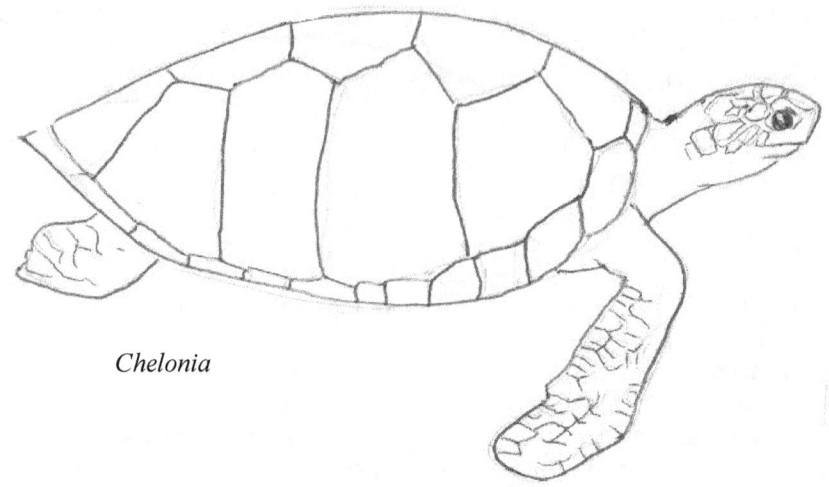

Chelonia

The green turtle *Chelonia myda* is the most herbivorous of the living turtles. Adults feed almost exclusively on sea grasses although juveniles are carnivorous, feeding on marine invertebrates and fish. As with the hawksbills in the Carettini, greens are almost entirely tropical and rarely nest outside of 0-25° latitude. The green turtle has three or four subspecies: *C. m. mydas* (Atlantic and Mediterranean), *C. m. agassizii* (Pacific Ocean), *C. m. japonica* (Indian Ocean) and *C. m. carrinegra* (north-east Pacific Ocean). The status of the latter is unclear as it may not be distinct from the main Pacific population, and at least 16 further names have been applied to different populations at one time or another. The 'black sea turtle' *C. m. agassizii* was described as distinct in 1868 as *Chelonia agassizii*, but is genetically indistinguishable from other green turtles and is generally considered to be the eastern Pacific Ocean subspecies. These black sea turtles are known from the Galapagos Islands to the coast of Mexico. They seem to differ from typical green turtles in their darker colour, smaller size and much longer tails in males. Some differences in skull shape have been reported but are not very clear. In genetic studies black sea turtles from Mexico and the Galapagos are very similar and are related to green turtles from Hawaii and to those from Oman in the Persian Gulf. This is exactly the east-west pattern of relationship that would be expected from a single species over a large area.

Green turtles are endangered due primarily to human hunting, but also due to loss of nesting beaches. Green turtles have been consumed in vast numbers throughout much of human history; thousands were killed every year in the 19th and 20th centuries for turtle soup industries in Europe (particularly Britain) in addition to uncounted numbers being killed for meat. High levels of exploitation seem to have occurred in the distant past, at least in some localities. There are Bronze Age sites from around 2600BC in Oman, Bahrain and Abu Dhabi that contain the butchered remains of thousands of turtles. Historical records describe trade in eggs and scutes from Mesopotamia in 2064-2062BC. Later, the Romans were heavy users of tortoiseshell.

Natator is the most restricted of all sea turtles; it is limited to the flatback turtle *Natator depressus* which is found only around the coasts of Australia and New Guinea. Although related to the ocean going green turtle the flatback is a coastal species, rarely venturing into deep water. It also differs from the green turtle in being carnivorous, feeding mainly on jellyfish, sea cucumbers and bivalves. The strangest feature of this turtle is the keratinised layer of skin that covers the shell. Having skin with a rich blood supply overlying the shell is unusual in turtles, only being seen in the soft-shelled turtles where it allows them to use the body surface for gas exchange, enabling them to live in poorly oxygenated swamps without needing to rise to the surface to breathe very often. Flatback turtle hatchlings live in mangrove swamps where their flattened shape allows them to hide and the rich blood supply to the skin allows a similar life to that of soft-shelled turtles. The unusual shell may have evolved as a juvenile adaptation to life in swamps, and is retained in the adults. This may also explain the (   )m a more typical green turtle-like ancestor but adapted to a life in association with the mangroves, this would have separated it from the oceanic green turtles and led to its preference of shallow coastal waters and isolation around Australia.

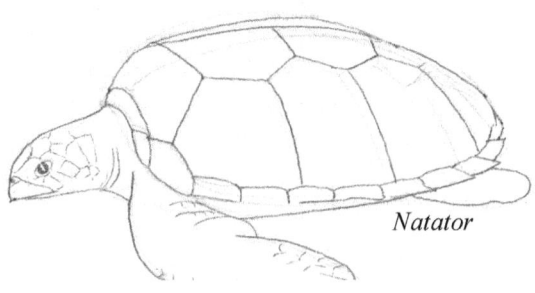

Natator

The evolution of the sea turtles has been strongly influenced by changes in ocean connections. The first major event may have been the closure of the Tethys Sea between Europe and Africa around 18 million years ago, which would have prevented the movement of turtles between the Indian and Atlantic Oceans north of Africa. The development of cold water currents in the southern oceans will have restricted turtle movement in southern regions from 17 million years ago to around 6 million years ago when current temperatures increased slightly. However, tropical migrations between the oceans would have been possible until the Pacific and Atlantic Oceans were separated around 3 million years ago when the Isthmus of Panama formed. Cold southern oceans prevent turtles moving between these oceans south of Cape Horn. This divided turtle populations in the two oceans, resulting in distinct Atlantic and Indo-Pacific subspecies in the green, hawksbill and loggerhead turtles. Green turtle populations are estimated to have been separated for the past 7 million years (estimates vary from 2 to 13 million years ago), suggesting that movement of green turtles between the oceans was reduced before the Isthmus closed finally. In the loggerheads the isolation led to speciation, with the olive ridley in the Indian Ocean and Kemp's ridley in the Atlantic. The olive ridley seems to have expanded its range from the Indian Ocean, retaining most genetic variation around the coast of India, with 96% of olive ridleys nesting in Orissa having a distinct mitochondrial DNA sequence. This species shows some genetic differences between Atlantic and Indo-Pacific populations, but this is relatively minor. There is the possibility of some movement between the Indian Ocean and the Atlantic via the connection south of the Cape of Good Hope. 5 million years ago (or possibly as early as 10 million years ago) the present current systems around southern Africa developed, with the warm Arghullas Current from the Indian Ocean colliding with the cold Atlantic Benguela Current, this would form a partial barrier to movement around the Cape. In the last 100,000 years these current patterns were disrupted and warm water flowed from the Indian Ocean into the Atlantic, allowing easy colonisation of the Atlantic for a brief period. Since then movement has again become difficult but occasional warm water gyres from the Arghullas Current move from southern Africa, round the Cape into the Atlantic. This seems to allow some loggerhead turtles to move into the Atlantic, giving a low level of gene flow in this direction but resulting in a close genetic connection between Indian Ocean and west Atlantic loggerheads. This route is also presumed to have been used by olive ridleys colonising the Atlantic from the Indian Ocean. East Atlantic hawksbills also seem to have a recent origin from the Indian Ocean. In green turtles there seems to have been very little gene flow between the Indian Ocean and the Atlantic over the past 1.5 million years. Turtles of the Mediterranean are of recent origin, having colonised since the flooding of the Mediterranean in the 'Zanclean Flood' 5 million years ago. Prior to that date the Mediterranean was a low-lying marshy depression isolated from the Atlantic. 5.3 million years ago the sea broke through the present day Straits of Gibraltar, flooding the depression. It seems that olive ridleys move between the oceans with considerably more difficulty than do most of the other species. This probably arises from differences in their dispersal patterns. After reaching the sea, hatchling sea turtles are thought to drift on ocean currents, spending several years in pelagic nursery areas. The exact location and movements of these nurseries are speculatory at present; larger juveniles have been fitted with satellite transmitters and tracked on journeys throughout the oceans. The older juveniles leave the oceanic waters and join adults in more coastal environments; in the case of most green, hawksbill and loggerhead populations the feeding and breeding areas are largely different and they migrate between these areas, sometimes swimming against the prevailing currents.

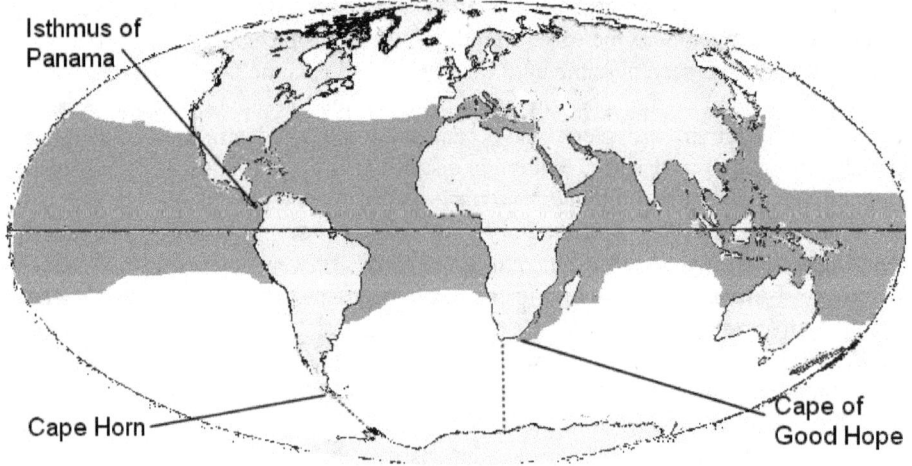

Barriers to sea turtle movement.

In contrast olive ridleys feed pelagically, drifting with currents rather than swimming against them on deliberate migrations. As the currents rarely cross from the Indian Ocean to the Atlantic, this movement will be much rarer in olive ridleys than in other species.

Atlantic green turtles show very little genetic structuring, what little structure there is has been shaped by geology, sea-level changes and recent exploitation. The Atlantic nesting populations seem to have diverged within the past 470,000 years, with the most studied rookery, Ascension Island, having been occupied only for the 88,000 years. The recency of the origin of this rookery may relate to sea-level changes; for most of the past 5 million years until around 100,000 years ago sea-levels were much lower than at present and volcanic islands such as Ascension would have represented steeply shelving islands with little or no beach. Even today Ascension offers only a small area for nesting. It seems that the only Atlantic island with an older population is Sao Tome, which conversely may have had a larger nesting beach area in the past due to the shallow water between it and West Africa. The pattern of genetic variation found in this population fits with the idea that they were nesting on the island more than 500,000 years ago and the population expanded rapidly 35,000 years ago, when the land currently 80 m below the surface of the sea would have been shallow water and beaches. Sao Tome does not seem to have been settled by humans until the 1470s, so would have been the most important green turtle nesting site in the Atlantic until comparatively recently. In contrast Bioko's turtles show very little genetic variation and may have started nesting on the island only 8,000 years ago. This island is separated from the African coast by a narrow stretch of shallow water, which would have been completely dry land 20,000 years ago. This recent origin population has also been affected by heavy levels of harvesting by humans for at least the last 3,000 years.

In the Indian Ocean green turtles are more diverse genetically than in the Atlantic, due to longer occupation of nesting islands; for example the 5,000 females nesting in the Comoros islands seem to have been breeding there for the past 182,000 years. In the Mozambique Channel there is a notable level of genetic variation with a major difference between populations breeding in the north and south of the channel. The oceanic circulation in the channel serves to minimise mixing between the north and south ends. There is a further subdivision in the south with a distinction between the Europa and Juan de Nova island populations. There is a partial barrier between the Indian and Pacific Oceans but as this is in the form of the Indonesian and Australasian islands this is easily traversed by turtles. As a result there is little if any difference between turtles in these oceans. In the Pacific there are no barriers, although the distances between the most easterly and westerly populations are considerable. In the loggerhead the east and west Pacific populations diverged within the last 250,000 years if at all. This may indicate that the east Pacific has been tropical for only a short time; cold upwelling during the Ice Ages may have made conditions too cold for tropical marine species.

References

Bourjea, J., S. Lapegue, L. Gagnevin, D. Broderick, J .A. Mortimer, S. Ciccione, D. Roos, C. Taquet & H. Grizel. 2007. Phylogeography of the green turtle, *Chelonia mydas*, in the Southwest Indian Ocean. *Mol. Ecol.* **16**: 175–186

Bowen, B.W. & S.A. Karl. 2007. Population genetics and phylogeography of sea turtles. *Mol. Ecol.* **16**: 4886–4907

Brinkman, D., M.C. Aquillon-Martinez, C.A. De Leon Dávila, H. Jamniczky, D.A. Eberth & M. Colbert. 2009. *Euclastes coahuilaensis* sp. nov., a basal cheloniid turtle from the late Campanian Cerro del Pueblo Formation of Coahuila State, Mexico. *PaleoBios* **28**

Danilov, I.G., A.O. Averianov & A.A. Yarkov. 2010. *Itilochelys rasstrigin* gen. et sp. nov, a new hard-shelled sea turtle (Cheloniidae *sensu lato*) from the lower Paleoene of Volvograd. *Proc. Zool. Inst. RAS* **314**: 24-41

Formia, A. B.J. Godley, J.-F. Dontaine & M.W. Bruford. 2006. Mitochondrial DNA diversity and phylogeography of endangered green turtle (*Chelonia mydas*) populations in Africa. *Conserv. Genet.* **7**: 353-369

Hirayama, R. 1994. Phylogenetic systematics of chelonioid sea turtles. *Isl. Arc.* **3**: 270–284

Hirayama, R. 1998 Oldest known sea turtle. *Appl. Opt.* **392**: 705–708

Hirayama R. & T. Chitoku. 1996. Family Dermochelyidae (Superfamily Chelonioidea) from the Upper Cretaceous of North Japan. *Trans. proc. Palaeont. Soc. Japan. NS* **184**: 597-622

Hirayama, R. & Y. Hikida. 1998. Mesodermochelys (Testudines; Chelonioidea; Dermochelyidae) from the Late Cretaceous of Nakagawa-cho, Hokkaido, North Japan. *Bull. Nakagawa Mus. Nat. Hist.* **1**: 69-76

Iverson, J.B., R.M. Brown, T.S. Akre, T.J. Near, M. Le, R.C. Thomson & D.E. Starkey. 2007. In search of the tree of life for turtles. *Chel. Res. Monogr.* **4**: 85–106

Jalil, N.-E., F. de Lapparent de Broin, N. Bardet, R. Vacant, B. Bouya, M. Amaghzaz & S. Meslouh. 2009. *Euclastes acutirostris*, a new species of littoral turtle (Cryptodira, Cheloniidae) from the Palaeocene phosphates of Morocco (Oulad Abdoun Basin, Danian-Thanetian). *Comptes Rendus Palevol* **8**: 447-459

Karl, S.A. & B.W. Bowen. 1999. Evolutionary Significant Units versus geopolitical taxonomy: molecular systematics of an endangered sea turtle (genus *Chelonia*). *Conserv. Biol.* **13**: 990-999

Kear, B.P. & M.S.Y. Lee. 2005. A primitive protostegid from Australia and early sea turtle evolution. *Biol. Lett.* doi: 10.1098/rsbl.2005.0406.

Lehman, T.H. & S.L. Tomlinson. 2004. *Terlinguachelys fischbecki*, a new genus and species of sea turtle (Chelonioidea: Protostegidae) from the Upper Cretaceous of Texas. *J. Paleont.* **78**(6): 1163–1178

Luschi, P., G.C. Hays & F. Papi. 2003. A review of long-distance movements by marine turtles, and the possible role of ocean currents. *Oikos* **103**: 293-302.

Lynch, S. & J.F. Parham. 2003. The first report of hard-shelled sea turtles (Cheloniidae *sensu lato*) from the Miocene of California, including a new species (*Euclastes hutchisoni*) with unusually plesiomorphic characters. *PaleoBios* **23**(3): 21–35

Mosseri-Mailio, C. 2000. The ancient distribution of sea turtle nesting beaches: an archaeological perspective from Arabia. *Testudo* **5**(2): 31-36

Naro-Maciel, E., M. Le, N.N. FitzSimmons & G. Amato. 2008. Evolutionary relationships of marine turtles: A molecular phylogeny based on nuclear and mitochondrial genes. *Mol. Phyl. Evol.* **49**: 659-662

Parham, J.F. & N.D. Pyenson. 2010. New sea turtle from the Miocene of Peru and the iterative evolution of feeding ecomorphologies since the Cretaceous. *Journ. Paleontol.* **84**(2): 231-247

Parham, J.F. & T.A. Stidham. 1999. Late Cretaceous sea turtles from the Chico Formation of California. *PaleoBios* **19**(3): 1-7

Shanker, K., J. Ramadevi, B.C. Choudhury, L. Singh & R.K. Aggarwal. 2004. Phylogeography of olive ridley turtles (*Lepidochelys olivacea*) on the east coast of India: implications for conservation theory. *Mol. Ecol.* **13**: 1899–1909

Weems, R.E. 1974. Middle Miocene sea turtles (*Syllomus, Procolpochelys, Psephophorus*) from the Calvert Formation. *Journ. Paleontol.* **48**: 278-303

Wood, R.C., J. Johnson-Gove, E.S. Gaffney & K.F. Maley. 1996. Evolution and phylogeny of leatherback turtles (Dermochelyidae), with descriptions of new fossil taxa. *Chelonian Conserv. Biol.* **2**: 266-286

Zangerl, R. & W.D. Turnbull. 1955. *Procolpochelys grandaeva* (Leidy), an early Carettine Sea Turtle *Fieldiana Zoology* **37**: 345-384

9. Swamp monsters – mud, stink and snapping

The relationships between snapping turtles (Chelydridae), mud turtles (Kinosternoidea) and the sea turtles (Chelonioidea) are still not clear. Sea turtles are probably most closely related to the tortoise and terrapin group Testudinoidea but as they are ecologically quite different from other turtles, the sea turtles are treated separately. The snappers and the mud turtles are considered below. Chelydridae and Kinosternoidea almost certainly share a North American origin and are limited to the Americas today, although chelydrids did once live in Europe as well. There are two equally likely relationships between these families: either Kinosternidae and Chelydridae are sister families, separate from the related Chelonioidea and Testudionoidea, or the Kinosternoidea are the most primitive of the four, followed by the Chelydridae.

The family Chelydridae includes the snapping turtles *Chelydra* and *Macroclemys*. The big-headed turtle *Platysternon* is sometimes thought to be related to

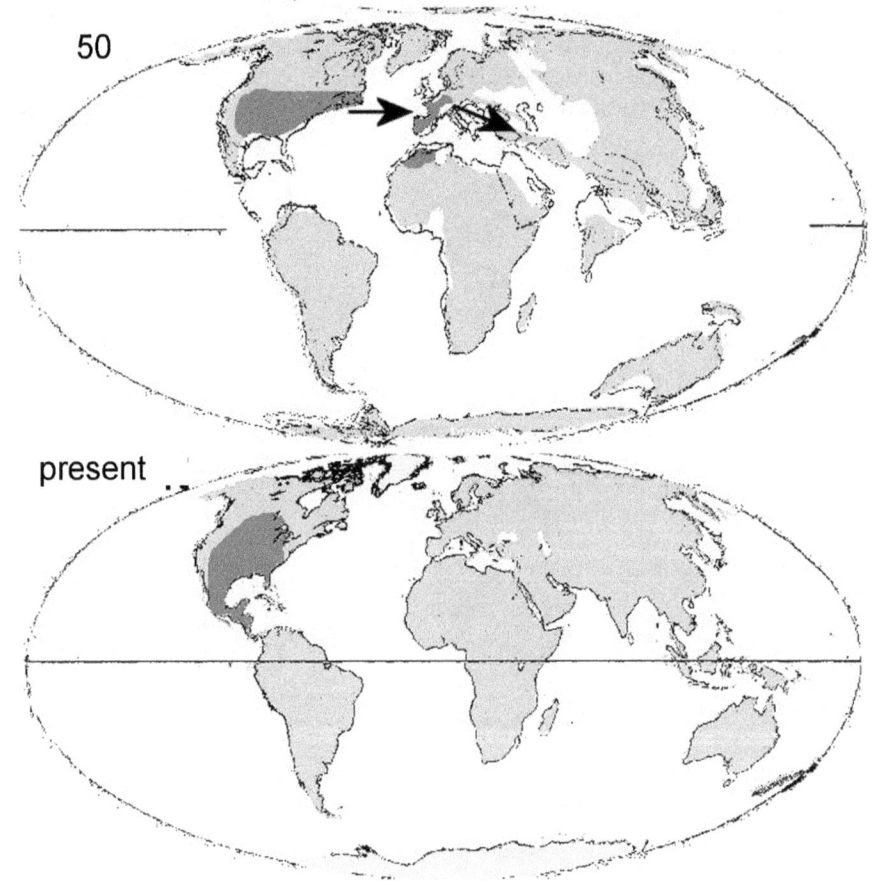

Distribution of Chelydridae from 50 million years ago.

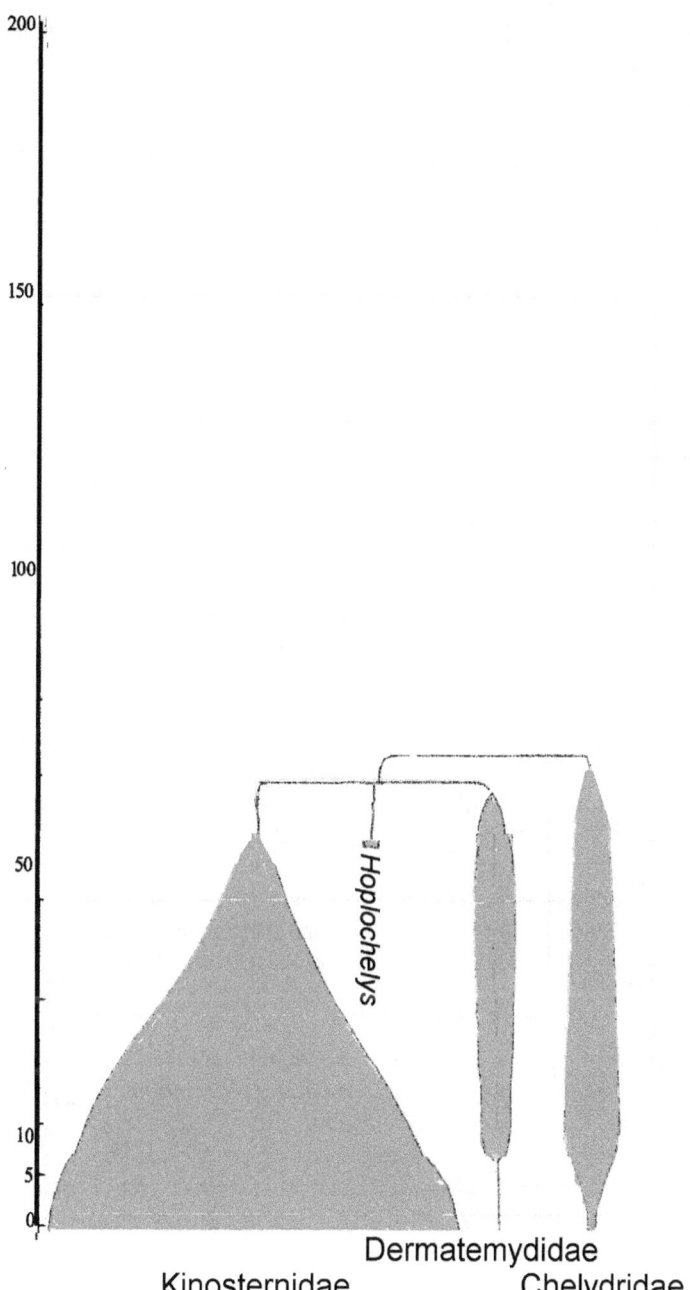

Phylogeny of kinosternids, dermatemydids and chelydrids.

the snappers but this is probably simply due to their somewhat similar appearance. Molecular studies increasingly suggest that the big-headed turtle is related to the testudinids, and to the emydid terrapins in particular. The earliest relatives of the living Chelydridae were the panchelydrids from 90 million years ago of North America, the Chelydridae themselves are known from around 60 million years ago. Chelydridae seem to be related to the Dermatemydidae and Kinosternidae. The three groups appear to have evolved at more or less the same time (within the space of a million years), and the relationships between them are far from clear.

The Chelydridae appear to have a long fossil history dating back to about 60 million years ago, with forms such as *Acherontemys hehmanni* and *Chelydrops*. *Hoplochelys* has been thought to be a chelydrid genus but is probably a dermatemydid. One species referred to the genus, *H. caelata*, from 55 million years ago in North America does seem to be a chelydrid but is probably not a true *Hoplochelys*. *Chelydrops* is known from a single species, *C. stricta* from 20 million years ago in North America. The oldest known possible chelydrid is *Emarginachelys cretacea* 80 million years ago from North America, but this has also been considered to be a trionychid or kinosternoid. One well preserved species is *Protochelydra zangerli* from North Dakota 60 million years ago. It has a higher domed carapace than living snapping turtles; this strong, domed shell has been suggested to be associated with the presence of large turtle-eating crocodilians. It has been placed in its own subfamily, Protochelydrinae. It has a more emarginate skull than in the Chelydrinae, the jaws are relatively wide and there is a ridge on the ventral surface of the pterygoid with projects posteriorly, associated with powerful jaw muscles. As in Chelydrinae the plastron is reduced to a small, cross-shaped structure but lacks fontanelles.

True snapping turtles (Chelydrinae) are, as their name implies, highly effective ambush predators. They are generally slow moving, lurking on the bottom of water bodies and catching prey (mainly fish) with a quick bite. The ancestors of the family were North American (panchelydrids) and the family is now entirely restricted to North and Central America. However, formerly it was found throughout much of Europe. Around 50 million years ago they were present in Europe and North Africa (*Gafsachelys*), but were never diverse in that region. These animals were able to cross the narrow and shallow channels between North America, Greenland and Europe around 60 million years ago. The crossing points would have been at high latitudes, with cold climates but some of the living snapping turtles are able to tolerate cold water, being found as far north as central Canada. The European species included the 75 cm long *Chelydropsis* from 34 to 5 million years ago of Europe and Asia and *Macrocephalochelys* from Ukraine 5 million years ago. This latter population was found in the rivers and swamps around the shores of what was a considerably larger Black Sea 5 million years ago. 4 million years ago Ice Age conditions affected Europe and Asia; ice moved south over much of the continent, forcing turtle populations south. The snapping turtles did not survive these changes and for the past 4 million years have been restricted to the Americas.

Of the living species there is a deep divergence between *Chelydra* and *Macroclemys*. The former has three living species and the latter only one. The North American or common snapping turtle *Chelydra serpentina* is widespread throughout

Chelydra

much of North America and is replaced by two relatively narrow range species in Mexico: the South American snapper *C. acutirostiris* and the Central American snapper *C. rossignoni*. Within *C. serpentaria* there are two subspecies: *C. s. serpentina* and *C. s. osceola*. These are morphologically similar and almost identical genetically, suggesting a very recent and rather limited separation of the east and west populations. There is very weak differenation between species and a general lack of variation within *Chelydra*. All snapping turtles are highly aggressive predators, feeding on almost anything small enough to swallow, from algae, water weed, sponges and invertebrates to birds, other turtles and small mammals. They are highly adaptable, able to live in a wide variety of water bodies and laying up to 83 eggs per nest. This allows the common snapper to remain a common species in parts of its range despite hunting for food and consumption of eggs. Some populations have declined because of over-hunting and also pollution.

The alligator snapping turtle genus *Macroclemys* includes two early species: *M. schmidti* from 23 to 5 million years ago and *M. auftenbergi* from 5 million years ago. These may have declined in the Ice Age of 4-2.5 million years ago, the only alligator snapper surviving those cold conditions being the living alligator snapper *Macroclemys temminckii*. The alligator snapper has three distinct genetic groups: the Suwannee River population and more closely related western and central populations. The genetic differences are matched by differences in skull shape which changes between the western Pensacola Bay and the central Choctawhatchee river drainages. These areas are separated by a large sand-hill region. The genetically distinct Suwannee population is not morphologically distinctive but may be a remnant of the populations isolated in refugia during the Ice Age. This Suwannee relict may have survived because the cold waters

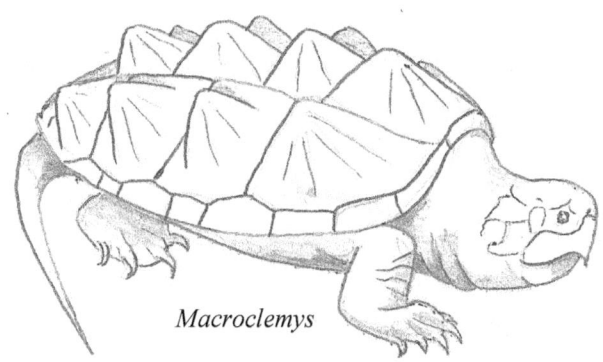

Macroclemys

101

running of glaciers would have had little impact on this southern river, unlike most of the prehistoric range of the species. Genetically this population is the closest alligator snapper to the common snapper.

The alligator snapper is a highly efficient predator, luring fish into its mouth by moving its worm-like tongue. Its general biology is similar to that of common snapping turtles but it is a considerably larger and more dangerous animal. Despite this it is occasionally kept in captivity. When animals become too large or unmanageable they are sometimes released into the wild, not only in their natural range of the south-central states of the USA, but also in France and Germany.

Fossils supposed to belong to the Dermatemydidae cover a wide time range and are found in Europe and North America; however, of the many fossil 'Dermatemydidae' only *Baptemys* and *Agomphus* from 56 to 34 million years ago in North America can be confidently placed in the Dermatemydae. Many fossil 'Dermatemydidae' may be closer to the Kinosternidae than to *Dermatemys*. The possible dermatemydid *Hoplochelys* was followed by the earliest true dermatemydid *Agomphus*. The closest genus to the living *Dermatemys* was *Baptemys*. *Hoplochelys* contains several species ranging from 65 to 60 million years ago, but the true number is difficult to determine due to the poor nature of the material. *Agomphus* of 65 million years ago of North America comprises several poorly known forms, most of which can be placed in a single species, *A. pectoralis*. These small (30 cm) possibly terrestrial turtles survived until 65 million years ago in North America. *Baptemys* comprises at least two distinct forms, but again has numerous poorly defined 'species'. These are more recent, covering 53 to 48 million years ago. Living Dermatemydidae are restricted to a single species, the Central American river turtle *Dermatemys mawii* of southern Mexico, Belize and northern Guatemala. This

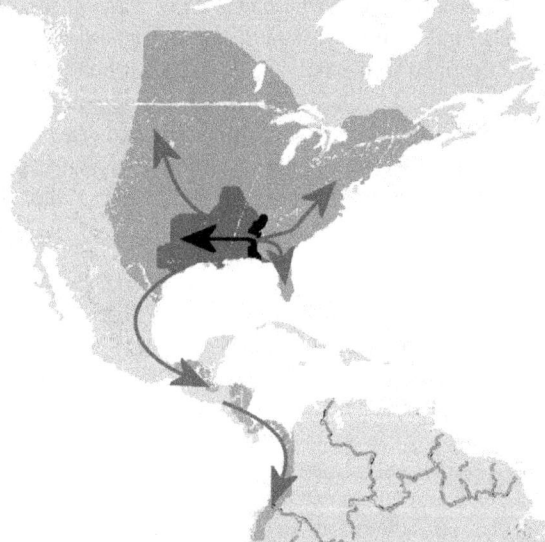

Distribution of living snapping turtles and dispersal patterns over the past 2 million years. *Chelydra* in light grey, *Macrochelys* in dark grey. Ancestral Suwanee River *Macrochelys* populations shown in black.

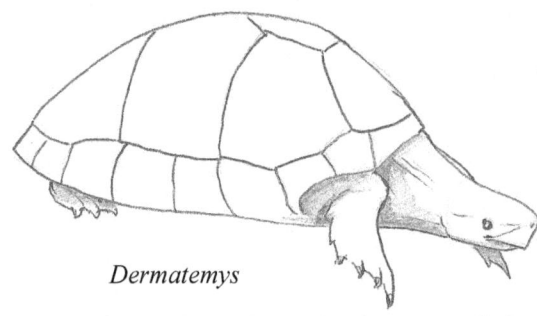

Dermatemys

is a relatively large fresh-water turtle at 65 cm long. It is highly specialised for an aquatic life and moves awkwardly on land, its paddle-like limbs being poorly adapted for walking and its neck muscles unable to support the weight of its head for long out of the water. Normally only females venture onto land for nesting. The turtles have a well-developed ability to absorb oxygen through their throats and rarely need to surface. Adults are vegetarian but juveniles are generally omnivorous. As a result of its large size it is heavily consumed by humans and populations are declining.

The oldest possible kinosternids date from 80 million years ago in North America but the first definite members of the family date from about 50 million years ago: *Xenochelys* and *Baltemys staurogastros*, again from North America. Living kinosternids are all American but range from the middle of North America down to the middle of South America. Living species can be divided into two subfamilies: the Staurotypinae and Kinosternoninae. Staurotypinae comprise the two highly distinctive musk turtle genera *Staurotypus* and *Claudius*. Both have disproportionately large heads with large jaws, a small shell with a remarkably small plastron and live in muddy waters. No fossil material has been ascribed to this subfamily.

Staurotypus are the giant musk turtles, comprising the northern giant musk turtle *Staurotypus triporcatus* in the Yucatan peninsula of Mexico, Belize and Guatemala, and the Pacific coast giant musk turtle *S. salvinii* in the southern part of the Yucatan. They are highly active and their large jaws make them formidable predators of a wide range of invertebrates (crustaceans, snails and insects) and fish; they may even eat smaller turtles from their sister subfamily the Kinosterninae. They feed mainly by suction, sucking prey into their mouths and then crushing them with their jaws. Despite their aggressive

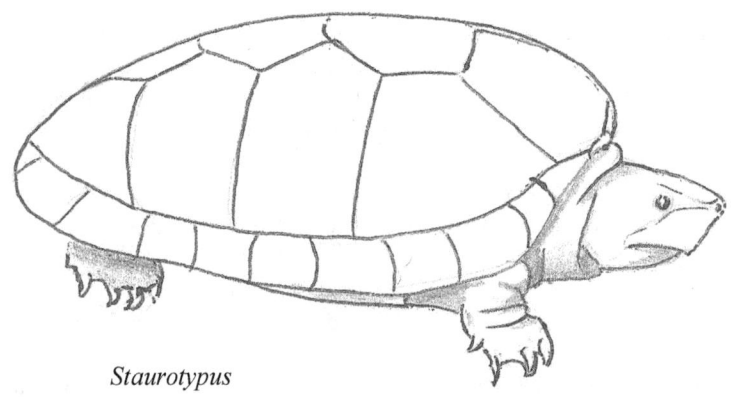

Staurotypus

demeanour they are hunted by people and their populations are in decline. The Pacific coast species has evolved a hinged plastron, allowing the front lobe to be drawn upwards. *Claudius* is a similarly ferocious genus containing just one species, the narrow-bridged musk turtle *Claudius angustatus*. This is found in southern Mexico into the Yucatan peninsula.

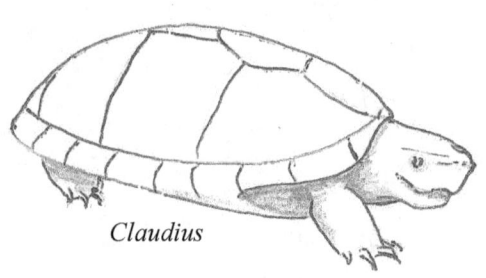

Claudius

Kinosterninae comprise the early *Baltemys staurogasto* from 53 million years ago, followed by *Xenochelys* from 53 to 35 million years ago. Of the two *Xenochelys* species *X. lostcabinensis* is the earliest, the more recent (35 million years ago) *X. formosa* may group with the living genera. Living Kinosterninae generally contains two genera, but *Sternothaerus* may be either the foundation of, or included within *Kinosternon*. *Sternothaerus* includes the widespread North American generalist common musk turtle or stinkpot *S. odoratum*, the south-east North American loggerhead musk turtle *S. minor* (a carnivore, specialising partly on molluscs with subspecies *S. m. minor* in the east and *S. m.*

Kinosternon

peltifer in the west of Alabama) and the flattened musk turtle *S. depressus* (a durophagous local endemic) and the south-western razor-backed musk turtle *S. carinatum* (an omnivore). The flattened musk turtle is highly aquatic, living in clear streams. As their name suggests, the musk turtles produce an unpleasant smelling secretion as a defence against predators. This is particularly apparent in the highly aggressive loggerhead musk turtle. It probably needs these attributes in order to survive against predators; at less than 14 cm long they should be vulnerable to many predators but in fact only alligators will willingly tackle an adult loggerhead. The species is under threat, but this is due to pollution and drainage of water bodies rather than hunting or predation.

The musk turtles probably evolved in Central America some 15 million years ago. There may have been an early divergence between a Central American ancestor of the loggerhead musk turtle and a form that invaded North America around 12 million years ago. This evolved into the common musk turtle and the flattened musk turtle. The loggerhead musk turtle may have moved into North America 10 million years ago, subsequently diverging into eastern (Florida) and western (Alabama) areas, resulting in the formation of the subspecies. The western *S. m. peltifer* subspecies would have been able to expand from its refuge through the connection of the Tennessee River system into Mobile Bay. Although the flattened musk turtle is restricted to the Black Warrior River within the range of *peltifer* it is genetically distinct and may be a relict of the earlier colonist species which has largely been displaced by *peltifer*.

True *Kinosternon* species have evolved a hinged plastron in common with several other genera of turtle. They are small semi-aquatic turtles and most are poor swimmers, usually walking along the bottom of water bodies rather than swimming. Within this genus there are two groups: those from Central America and the North American species.

North America has the Arizona mud turtle *K. arizonense*, striped *K. baurii*, yellow *K. flavescens* (extending into Central America) and eastern *K. subrubrum*. The eastern mud turtle is a very aquatic species found throughout eastern and southern coastal rivers. It divides into *K. s. subrubrum* from most of the USA, the Mississippi mud turtle *K. s. hippocrepis* from Arkansas to Texas and the Florida mud turtle *K. s. steindachneri* from Florida. The striped mud turtle is the least aquatic of the genus; it lives in shallow water and often moves over land between marshes. Central America has the most species: the Tabasco mud turtle *K. acutum*, Alamos *K. alamoseae*, narrow-bridged *K. angustipons*, Jalisco *K. chimalhuaca* (close to *oxaca* and *intergum*), Creaser's *K. creaseri*, Herrera's *K. herrerai*,

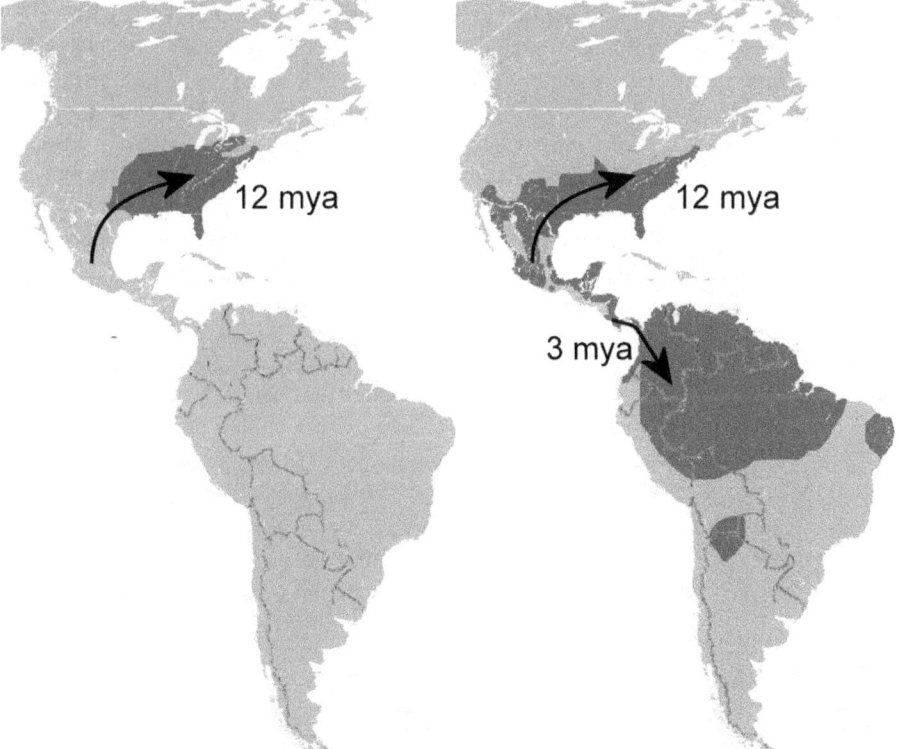

Distribution of *Sternothaerus* (left) and *Kinosternon* (right), showing timing of expansion from Central America.

rough-footed *K. hirtipes* (including some extinct subspecies), Mexican *K. integrum*, white-lipped *K. leucosteomum*, Oxaca *K. oxacae*, Durangoa *K. durangoense* (related to *flavescens*), and Sonora *K. sonoriensis*. Most of these live in temporary water bodies, although the white-lipped mud turtle is usually found in lakes and swamps. Despite this habitat preference it often ventures onto land. Even the Oxaca mud turtle favours temporary ponds although it lives in an area of permanent water courses. The Sonora mud turtle is probably the most aquatic of the Central American species, being found in slow forest streams of the Colorado and Yaquí River systems in northern Mexico and southern Arizona and California (where it is now extinct). It is divided into the main subspecies (*K. s. sonoriense*) and a highly isolated form found only on the Mexico/ Arizona border (*K. s. longifemorale*). The scorpion mud turtle *K. scorpiodes* extends from Central into South America. Its nominate subspecies is found in most of South America, with the central Chiapas *K. s. abaxillater* in Mexico, the white-throated *albogulare* from Colombia to Honduras and the red-cheeked from Mexico to Honduras. South America also has the isolated Dunn's mud turtle *K. dunni*. Of the Central American species the white-lipped and rough-footed mud turtles are the most widespread, and are divided into several localised subspecies. The white-lipped comprises the northern white-lipped mud turtle *K. l. leucostomum* and the southern *K. l. postinguinale*. The rough-footed includes the Valley of Mexico mud turtle *K. h. hirtipes*, Lake Chapala *K. h. chapalaense*, San Juanico *K. h. magdalense*, Viesca *K. h. megacephalum*, Mexican plateau *K. h. murrayi* and Patzcuaro *K. h. tarascense*. Of these the Viesca mud turtle from Coahuila province became extinct in about 1970. These diverse musk turtle are closely tied to their water courses but are not aquatic enough to cross large water bodies to move between different rivers. As a result they tend to be isolated in different river systems and so diversify into many similar species. This limited movement potential also makes them vulnerable to loss of small populations; as a result the extinction rate in the genus is unusually high.

References

Barley, A.J., P.Q. Spinks, R.C. Thomson & H.B. Shaffer. 2010. Fourteen nuclear genes provide phylogenetic resolution for difficult nodes in the turtle tree of life. *Mol. Phyl. Evol.* **55**: 1189-1194

Cervelli, M., M. Oliverio, A. Bellini, M. Bologna, F. Cecconi & P. Mariottini. 2003. Structural and sequence evolution of U17 small nucleolar RNA (snoRNA) and its phylogenetic congruence in chelonians. *Journ. Mol. Evol.* **57**: 73-84

Chandler, C.H. & F.J. Janzen. 2009. The phylogenetic position of the snapping turtles (Chelydridae) based on nucleotide sequence data. *Copeia* **2009**: 209–213

Feldman, C.R. & J.F. Parham. 2002. Molecular phylogenetics of emydine turtles: taxonomic revision and the evolution of shell kinesis. *Mol. Phyl. Evol.* **22**: 388–398

Gaffney, E.S. 1975. Phylogeny of the chelydrid turtles. *Fieldiana, Geology* **33**(9): 157-178

Gaffney, E.S. & P.A. Meylan. 1988. A phylogeny of turtles. *In* M.J. Benton (ed.) *The phylogeny and classification of tetrapods*. Oxford: Clarendon Press.

Hutchison, J.H. 1991. Early Eocene Kinosternidae (Reptilia: Testudines) and their phylogenetic significance. *Journ. Vert. Palaeont.* **11**: 145-167

Hutchison, J.H. & D.M. Bramble. 1981. Homology of the plastral scales of the Kinosternidae and related turtles. *Herpetologica* **37**: 73-85

Iverson, J.B., R.M. Brown, T.S. Akre, T.J. Near, M. Le, R.C. Thomson & D.E. Starkey. 2007. In search of the tree of life for turtles. *Chel. Res. Monogr.* **4**: 85–106

Knauss, G.E., W.G. Joyce, T.R. Lyson & D. Pearson. 2011. A new kinosternoid from the Late Cretaceous Hell Creek Formation of North Dakota and Montana and the origin of the *Dermatemys mawii* lineage. *Palaontol Z* **85**:125–142

Krenz, J.G., G.J.P. Naylor, H.B. Shaffer & F.J. Janzen. 2005. Molecular phylogeny and evolution of turtles. *Mol. Phyl. Evol.* **37**: 178-191

Meylan, P.A. & E.S. Gaffney. 1989. The skeletal morphology of the Cretaceous cryptodiran turtle, *Adocus*, and the relationships of the Trionychoidea. *Amer. Mus.Novit.* **2941**: 1-60

Meylan, P.A., R.T.J. Moody, C.A. Walker & S.D. Chapman. 2000. *Sandownia harrisi*, a highly derived trionychoid turtle (Testudines: Cryptodira) from the early Cretaceous of the Isle of Wight, England. *Journ. Vert. Palaeont.* **20**: 522-532

Parham, J.F., C.R. Feldman & J.L. Boore. 2006. The complete mitochondrial genome of the enigmatic bigheaded turtle (*Platysternon*): description of unusual genomic features and the reconciliation of phylogenetic hypotheses based on mitochondrial and nuclear DNA. *BMC Evolutionary Biology* **6**(11): 1-11

Roman, J., S.D. Santhuff, P.E. Moler & B.W. Bowen. 1999. Population structure and cryptic evolutionary units in the alligator snapping turtle. *Conserv. Biol.* **13**: 135-142

Shaffer H.B., P. Meylan & M.L. McKnight. 1997. Tests of turtle phylogeny: Molecular, morphological, and paleontological approaches. *Syst. Biol.* **46**: 235-268

Shaffer, H.B., D.E. Starkey & M.K. Fujita. 2008. Molecular insights into the systematics of the snapping turtles (Chelydridae). In: R.J. Brooks, A.C. Steyermark & M.S. Finkler (eds.) *Biology of the Snapping Turtle* (*Chelydra serpentina*). Johns Hopkins University Press, Baltimore

Thomson, R.C. & H.B. Shaffer. 2010. Sparse Supermatrices for Phylogenetic Inference: Taxonomy, Alignment, Rogue Taxa, and the Phylogeny of Living Turtles. *Syst. Biol.* **59**(1): 42–58

Walker, D., G. Ortí & J.C. Avise. 1998. Phylogenetic distinctiveness of a threatened aquatic turtle (*Sternotherus depressus*). *Conserv. Biol.* **12**: 639-645

10. Tortoises and terrapins

The living tortoises and terrapins all belong to the superfamily Testudinoidea. They may also include the big-headed turtle *Platysternon* but the relationships between these groups are still uncertain. Some studies, including mitochondrial DNA place the big-headed turtle with the snapping turtles, Chelydridae. Nuclear genes seem to give stronger support for them being the sister group to the terrapins, Emydidae, and this seems the most probable relationship.

The 'Cryptoderinea' has been proposed to combine the testudinoids with *Platysternon* in an arrangement first suggested by Vaillant in 1894 based on the arrangement of the joints between the neck vertebrae, but this remains disputed. The only close relatives of *Platysternon* seem to be some fragmentary fossils from between 65 and 23 million years ago in Kazakhstan, which are placed with *Platysternon* in 'Panplatysternon'. Another early fossil genus, *Anhuichelys* from China, is a probable testudinoid but resembles Platysternidae and Chelydridae. It includes several species, dating from 65-60 million years ago but does little to clarify the muddle. These 'Cryptoderinea' seem to have originated as freshwater turtles in Asia around 100 million years ago, diverging into the mountain stream *Platysternon* and the more varied and generalist ancestral terrapins from 95 million years ago.

The only living species of *Platysternon* is the big-headed turtle *P. megacephalum*, found in central China and throughout south-east Asia. Thehere are three subspecies in this range: *P. m. megacephalum* in China, *P. m. peguense* in Burma and Thailand and *P. m. shiui* in Cambodia, Laos and Vietnam. It is a fairly small turtle, at less than 19 cm long, but is a fearsome animal due to its massive head; 'megacephalum' is highly appropriate as the head is more than a third of the shell length. The powerfully hooked jaws give it a ferocious bite. This is used to catch snails, crustaceans and fish. It is a poor swimmer, with legs developed for walking and climbing. Its preference is for climbing around on the bottom of rapid running streams. It is even able to climb up waterfalls using the long, very muscular tail as a prop and the massive head for leverage. Although it is aggressive and produces a foul-smelling defensive musk, it is often eaten by humans and is threatened by hunting and by damming of its rivers for hydroelectric schemes.

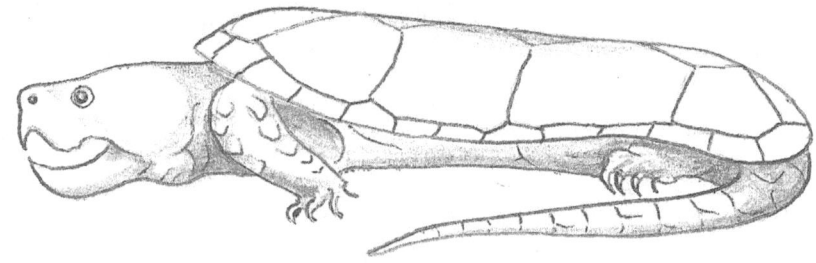

Platysternon

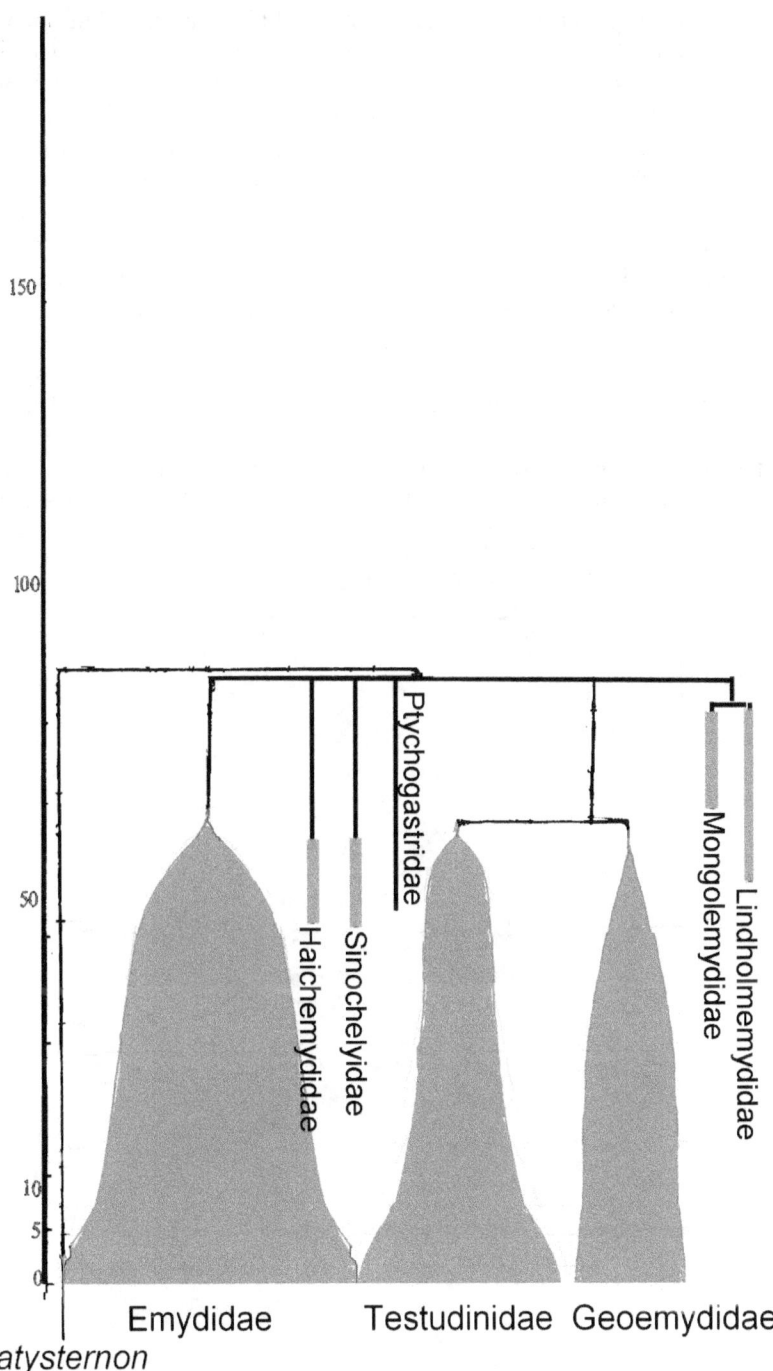

Phylogeny of Testudinoidea and *Platysternon*.

Whilst the big-headed turtle is of uncertain affinities, there is little doubt that the tortoises and terrapins belong together in the well-supported Testudinoidea. This comprises the tortoises (family Testudinidae) and the two largely fresh-water families Emydidae and Geoemydidae. Early testudinoids of uncertain affinities include the families Haichemydidae, Sinochelyidae, Lindholmemydidae and Mongolemydidae, most of which were found in either Mongolia or China, rarely both (the lindholmemydid *Gravemys* being one of the few genera found in both localities). In addition some Lindholmemydidae have been recorded from Uzbekistan (*Khodzhakulemys* about 80 million years ago) and Kazakhstan (*Linhdolmemys* 90 to 85 million years ago).

Haichemydidae contains only *Haichemys*. These were small (under 20 cm) turtles of Mongolia. Sinochelyidae was a more diverse family restricted to China (including *Heishanemys*, *Peishanemys* and *Sinochelys*); both families lived from 65 to 55 million years ago. The Mongolemydidae and Lindholmemydidae are closely related, with some genera such as *Khodzhakulemys* and *Elkemys* appearing to be intermediate between the two. The families were found in much the same region and at more-or-less the same time (mostly between 95 and 70 million years ago). The Mongolemydidae

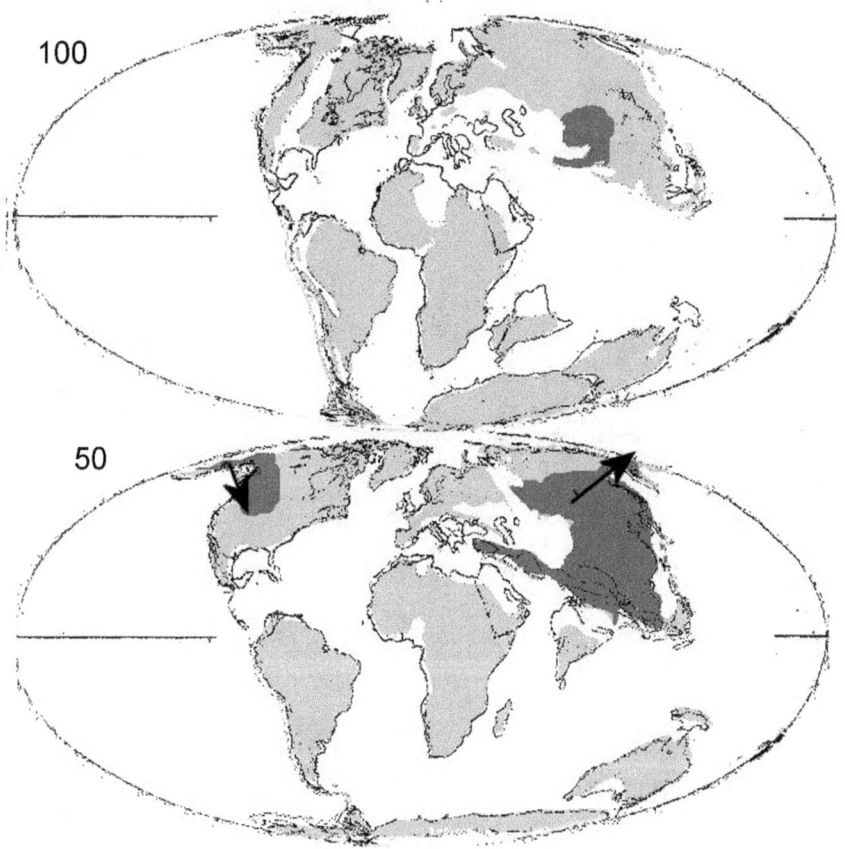

Distribution of early testudinoids from 100 to 50 million years ago.

were small, thin-shelled species (25 cm), largely associated with seasonal pools, whereas the Lindholmemydidae were mostly larger species of deeper water in large rivers or coastal environments. Inland environments favoured the Mongolemydidae where shallow seasonal pools were present. These habitat distinctions probably explain why the Mongolemydidae are Mongolian and the Lindholmemydidae largely Chinese as freshwater habitats in these geographical areas were quite different 95 million years ago; Mongolia was semi-arid with shallow and temporary water bodies, whilst China was more montane with higher rainfall and larger rivers flowing to the coast.

Mongolemys elegans (95 to 75 million years ago) shells often have tooth marks from the crocodile *Shamosuchus*. This was a strange crocodile with crushing teeth; it is assumed to have fed largely on molluscs but the damaged turtle shells indicate that as far as this crocodile was concerned, the difference between a clam and a turtle was unimportant. The presence of such turtle-eating crocodiles and the highly predatory ichthyodectid 'bulldog' fish may have limited the distribution and abundance of the Mongolemydidae, but apparently had little impact on the Lindholmemydidae. 93 million years ago there was a very rapid rise in sea level, the 'Turonian transgression' which flooded most low-lying environments in what is now central Asia and separating present day eastern

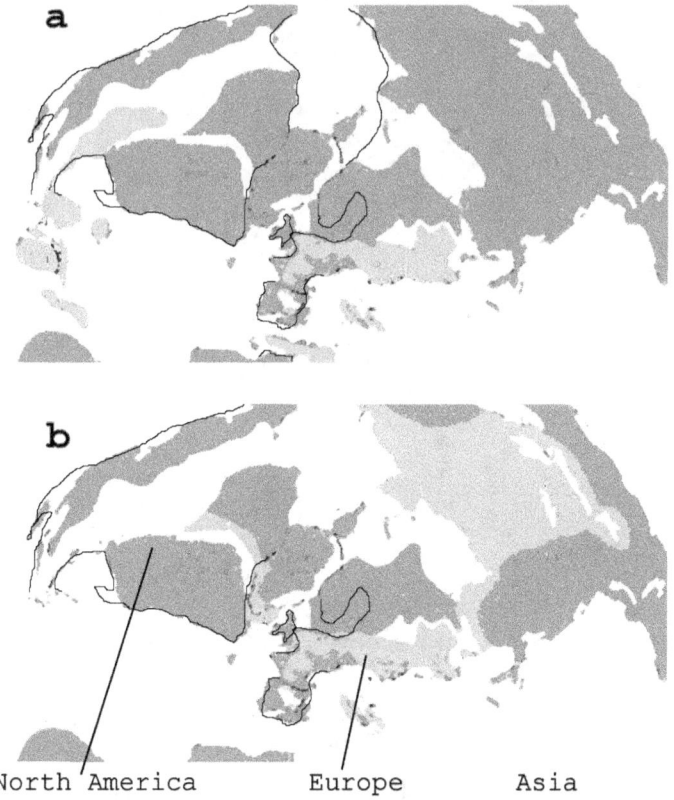

The Turonian Transgression: before (a) and after (b) - shallow water in light grey.

Asia from the fragmentary lands of Europe and west Asia. This was a major blow for the Mongolemydidae, facing an expansion of deeper marine waters and increased risk of predation from the crocodiles and giant fish. This turtle family declines from this point and disappears around 70 million years ago. Lindholmemydidae genera such as *Zangerlia* and *Gravemys* were more able to cope with this event and lasted until 55 million years ago.

The dramatic changes of the Turonian transgression left the northern hemisphere with few surviving freshwater turtles. The survivors included the remnants of the Lindholmemydidae in east Asia, soft-shelled turtles in Asia and North America, semi-aquatic kinosternids and snapping turtles in North America, and the Ptychogastridae in North America and Europe.

The oldest member of the Ptychogastridae was *Echmatemys* of North America (55 million years ago). These probably originated in Asia and crossed into North America via the Bering Land Bridge. This was a diverse genus in that continent

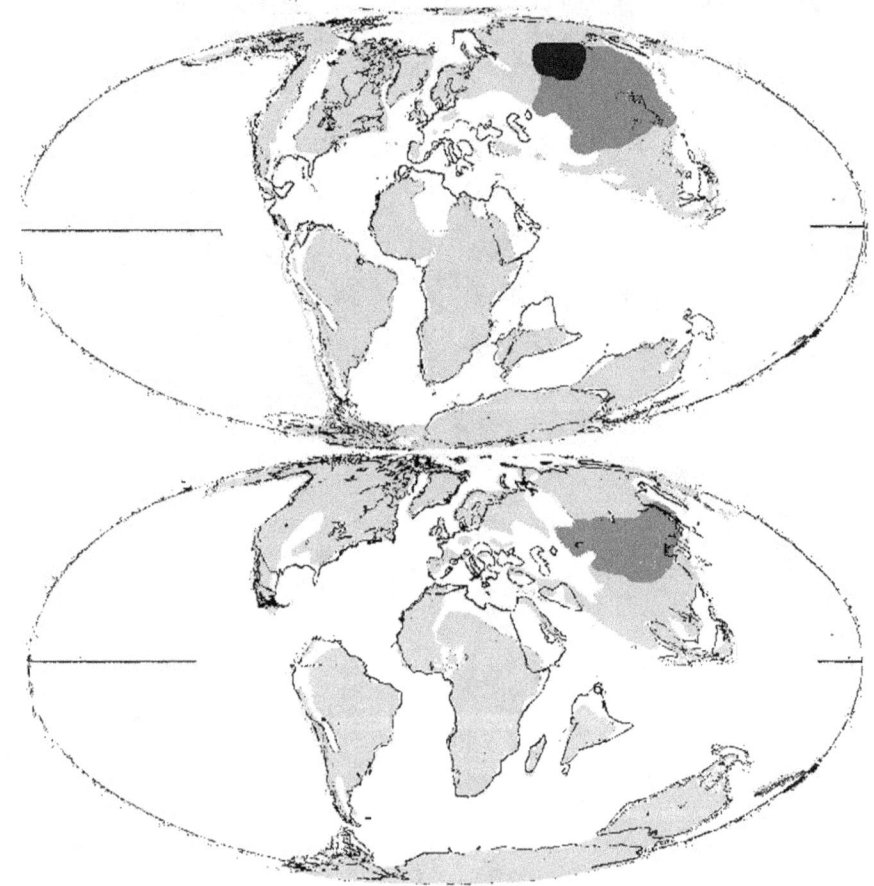

Distribution of Mongolemydidae (black) and Lindholmemydidae (dark grey) before and after the Turonian transgression (95-70 million years ago).

with several species known from between 55 and 37 million years ago. The founding European species was *Hummelemys*, followed by *Merovemys*, *Geiselemys* and *Ptychogaster* (including species referred to the dubious genus '*Temnoclemmys*'). In addition *Clemmydopsis* (15 to 8 million years ago) has been described from Europe but its relationship to the other genera is not known. These forms differ from one another in shell features, the most notable of which is the development of a plastral hinge in *Ptychogaster*. These ptychogastrids were replaced by the Emydidae in North America by 40 million

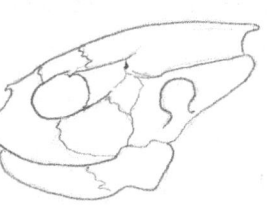

Mongolemys

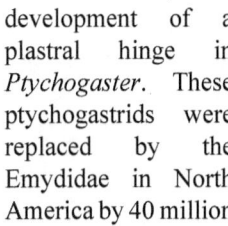
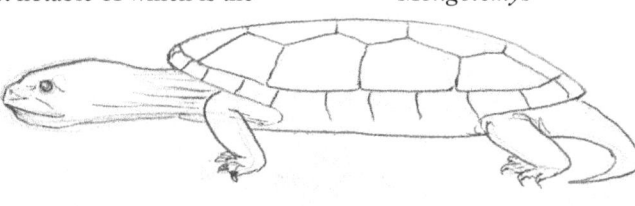

years ago and Europe by 10 million years ago. The extinction of most of the early freshwater turtles by around 50 million years ago left open the freshwater niche for the expansion of the American terrapins (Emydidae) and the Asian river turtles (Geoemydidae).

References

Auffenberg, W. 1974. Checklist of fossil land tortoises (Testudinidae). *Bull. Fl. State Mus., Biol. Sci.* **18**(3): 121-251

Barley, A. J., P. Q. Spinks, R. C. Thomson & H.B. Shaffer. 2010. Fourteen nuclear genes provide phylogenetic resolution for difficult nodes in the turtle tree of life. *Mol. Phyl. Evol.* **55**: 1189-1194

Hervet, S. 2003. *Le groupe* Palaeochelys *sensu lato*–Mauremys *dans le contexte systématique des Testudinoidea aquatiques du Tertiaire d'Europe occidentale.* Unpublished Ph.D. thesis, Muséum national d'Histoire naturelle, Paris.

Hervet, S. 2006. The oldest European ptychogasterid turtle (Testudinoidea) from the lowermost Eocene amber locality of Le Quesnoy (France, Ypresian, MP7). *Journ. Vert. Palaeont.* **26**: 839-848

Hutchison, H.J. 1980. Turtle stratigraphy of the Willwood Formation, Wyoming: preliminary results. *Paleontology* **24**: 115-118

Hutchison, H.J. 1996. Testudines. In: Prothero, D.R. & R.J. Emry (eds.). *The Terrestrial Eocene-Oligocene transition in North America.* Cambridge University Press

Mlynarski, M. 1976. Testudines. In *Handbuch der Palaeoherpetologie* 7

Parham, J.F., C.R. Feldman & J.L. Boore. 2006. The complete mitochondrial genome of the enigmatic bigheaded turtle (*Platysternon*). *BMC Evol. Biol.* **6**(11): 1-11

Wiens, J.J., C.A. Kuczynski & P.R. Stephens. 2010. Discordant mitochondrial and nuclear gene phylogenies in emydid turtles: implications for speciation and conservation. *Biol. Journ. Linn. Soc.* **99**: 445-461

11. Terrapins – Emydidae

The typical fresh-water terrapins of the family Emydidae appear to have originated some 55 million years ago in the northern hemisphere. The northern continents would have been highly fragmented at this time but separated only by narrow, largely shallow waterways; it is therefore probable that the emydids would have been able to spread throughout the northern continents no matter where they originated. A North American origin is suggested by the fact that living species are primarily North American and have been well established for at least 14 million years with fossils of the North American genera *Terrapene* and *Emydoidea*. There is a suggestion that the emydids evolved from the 55 million year old ptychogastrid *Echatemys* of North America. On biogeographical grounds this is plausible, and a 55 million year old *Echatemys* ancestor would be contemporaneous with the earliest North American emydids.

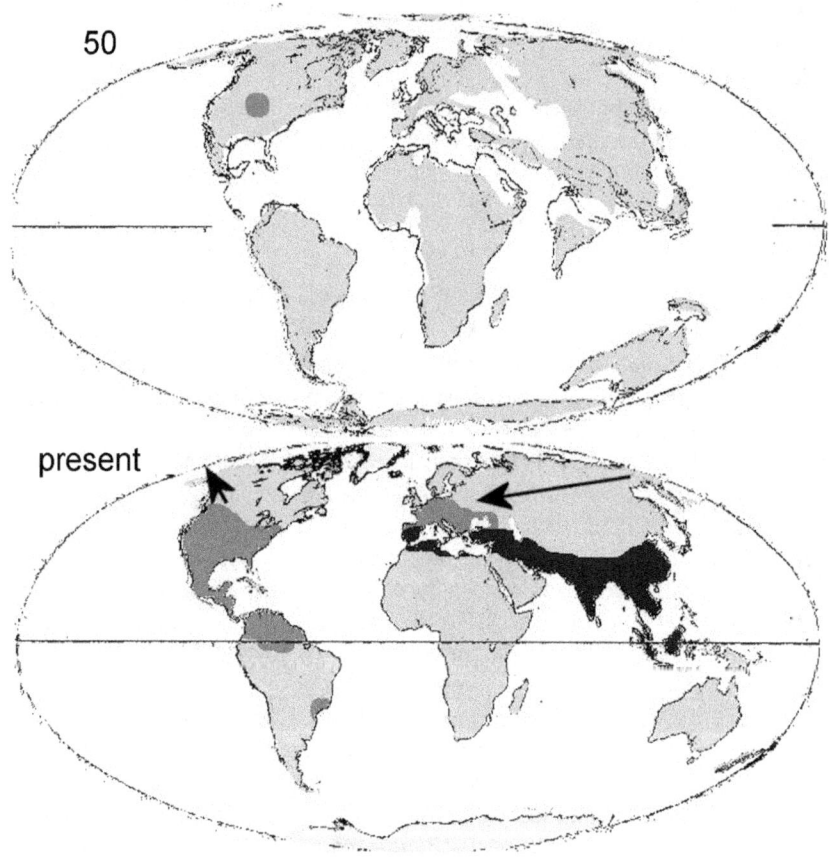

Distribution of early Emydidae from 50 million years ago, showing North American origins, dispersal westwards into Eurasia. Range of Eurasian Geoemydidae in black.

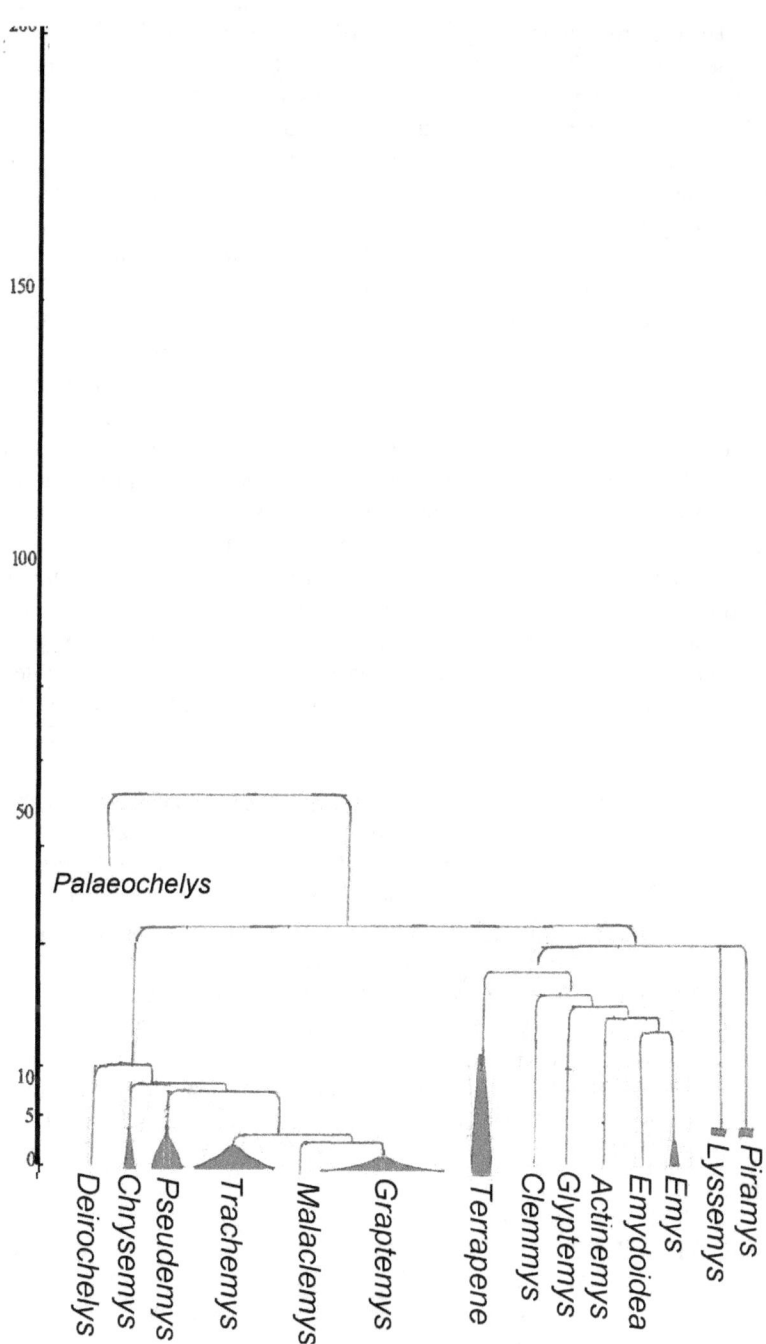

Phylogeny of Emydidae.

The earliest probable emydids are *Palaeochelys* (55 to 35 million years ago of North America). From North America it may have been possible for an early emydid to colonise Europe and East Asia. However, this is difficult to substantiate as early emydid fossils are rather nondescript and most remains identified as Emydidae are now placed in other families (mostly Geoemydidae, but also some Lindholmemydidae and Ptychogastridae). European fossils from as early as 34 million years ago have been considered emydids but these are either too fragmentary to identify properly or are now placed in other families. The suggested early east Asian emydid *Hokouchelys* (55 million years ago) may also be a misidentification of the geoemydid *Grayemys*. In the absence of reliable early Eurasian emydids the earliest non-American species may be the Indian fossils from 5 million years ago: *Lyssemys piramensis* and *Piramys auffenbergi*, although even there some Indian 'emydid' fossils from 5 million years ago have now been referred instead to the living geoemydid genus *Kachuga*. This rather unsatisfactory situation may be expected; if the emydids evolved in North America some 55 million years ago they would have been isolated there until 20 million years ago when falling sea-levels exposed the Bering Bridge between Alaska and Siberia. At this time emydids would have been able to colonise East Asia. They may have spread into Asia but been unable to become fully established in much of the continent due to competition from the already well established Asian Geoemeydidae. The colonising emydids would have only survived around the edges of the geoemydid range, in newly available lands following climate changes and geological movements. Suitable areas for their colonisation may have been the Indian subcontinent which contacted Asia some 40-50 million years ago (if *Lyssemys* and *Piramys* really are true emydids), and western Asia and Europe.

Living members of the family can be divided into two lineages: the main pond, bog and box turtle group *Glyptemys-Clemmys-Actinemys-Emys-Emydoidea-Terrapene* and the slider and cooter group *Deirochelys-Chrysemys-Pseudemys-Trachemys-Malaclemys-Graptemys*. Many DNA studies have found contradictions between the stories provided by nuclear and mitochondrial DNA in the Emydidae as a result of frequent movements between populations and hybridization. All studies do agree that the living species are descended from North American ancestors and the only genus found outside the Americas is *Emys* from Europe and western Asia. The living genera are identifiable in the fossil record from around 20 million years ago.

Based on mitochondrial genes the main living emydid lineage divides first between *Clemmys* and all other emydids. Then follows a division into three groups: one comprising just *Glyptemys* (sometimes referred to as *Calemys*), another of the Terrapene species and a third group including *Actinemys*, *Emys* and *Emydoidea*.

The spotted turtle *Clemmys guttata* seems to be the most isolated of all emydids in terms of its origins. This is a very distinctive species, with a black

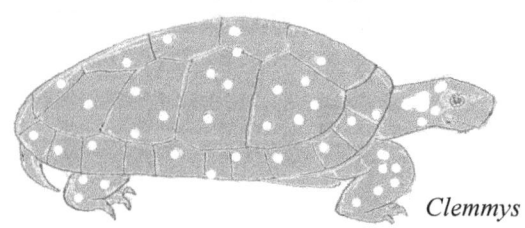

Clemmys

116

shell and body, covered in yellow spots. It is largely aquatic, living in shallow water bodies. It is widespread in eastern North America, being able to survive cold winters. They are small (up to 13 cm), with strongly ridged jaws. The gulars project to form a groove into which the head fits, this is a unique arrangement of the front of the shell. Although many fossils have been ascribed to this genus these are too poorly preserved or defined to be identifiable, and most are now placed in other genera.

Glyptemys arose around 15 million years ago with fossils of *G. valentinensis* from Nebraska. The living species are partially aquatic North American turtles. The

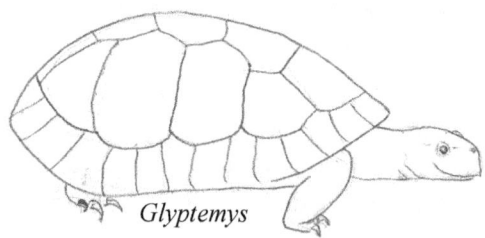
Glyptemys

wood turtle *G. insculpta* is found throughout North America, as far north as Quebec and is known from fossils from up to 2 million years ago to the present. The bog turtle *G. muhlenbergi* is widespread in the USA. As their name suggests, the wood turtle is found in forest habitats. It is largely terrestrial, although it bury itself in mud in summer. Similarly, the bog turtle is really a terrestrial species that lives in marshy habitat. Both are omnivorous and seem to have evolved as a result of habitat specialisation, one for forest and one for marshy open ground. The wood turtle is known from 5 million years ago to the present and the bog turtle from 500,000 years ago to the present. The fossil species *G. valentinensis* may be the last common ancestor of the two living species.

Members of the genus *Terrapene* are the box turtles from North and Central America. These small turtles have an aquatic origin but several forms have adapted to terrestrial life, filling the tortoise niche in much of their range. Most species have at least some mobility in the front and back lobes of the plastron, allowing the shell to be sealed completely. *Terrapene* were the first turtles to be called 'terrapins', this being derived from the Algonquin name for the animals in the north-east USA. Now they are called 'box turtles' in reference to the plastral hinges.

The genus is known from about 14 million years ago. The earliest remains are similar to the ornate box turtle *T. ornata* and are known from Nebraska. These deposits also include an extinct fossil species *T. corneri*. Fossils similar to the Eastern or

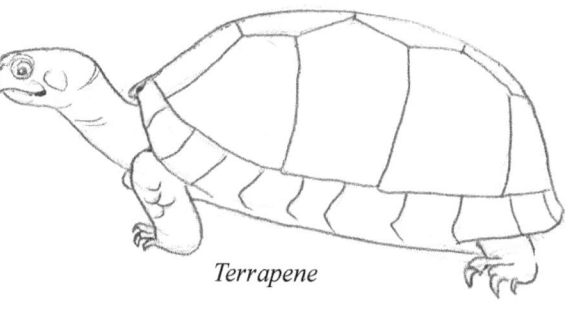
Terrapene

common box turtle *T. carolina* date from around 12 million years ago from Kansas. The ornate and common box turtles are the two most widespread of the living box turtle species and the only ones with overlapping ranges; the others are largely geographically isolated and more specialised in their habitat preferences. The common box turtle occurs from southern Canada to southern Mexico. It comprises the eastern box turtle *T. c. carolina*, the Florida box turtle *T. c. bauri* restricted to Florida (which may be regarded as a distinct species), the Gulf Coast box turtle *T. c. major* of coastal southern USA, Mexican box turtle *mexicana* from central Mexico, three-toed box turtle *T. c. triunguis* from western USA and the Yucatan box turtle *T. c. yucatana* from southern Mexico. Although the three-toed box turtle is named for the reduced toes in the hind feet, the Florida and Mexican box turtles also have only three toes.

The spotted box turtle is divided into geographically very isolated subspecies: the southern *T. n. nelsoni* (Nayarit and Sinaloa states) and northern *T. n. klauberi* (Sinaloa and Sonora states). The ornate box turtle contains the subspecies *T. o. ornata* from much of the USA and the desert box turtle *T. o. luteola* from Mexico and the southern USA. Most of the species are brightly coloured, with light brown to yellow patches on the shell, yellow or orange scales on the legs, neck and head and distinctively red irises in males of most species. They are quite adaptable in habits, and tend to be found in damp habitats although they are generally terrestrial. The most terrestrial are the spotted and desert box turtles which live in arid areas. All of them are able to swim and most will willingly enter water. The most aquatic is the Coahuilan box turtle which is restricted to the Cuatro Cinégas valley of Mexico. There they live in pools that are either completely fresh or highly brackish, due to the unusual geology of the area. Although this is a largely aquatic species it feeds out of water, suggesting that it has come from a more terrestrial, more typically *Terrapene* ancestor. It is also the least colourful box turtle, being generally a dull brown. Due to its extremely restricted distribution it is considered to be endangered.

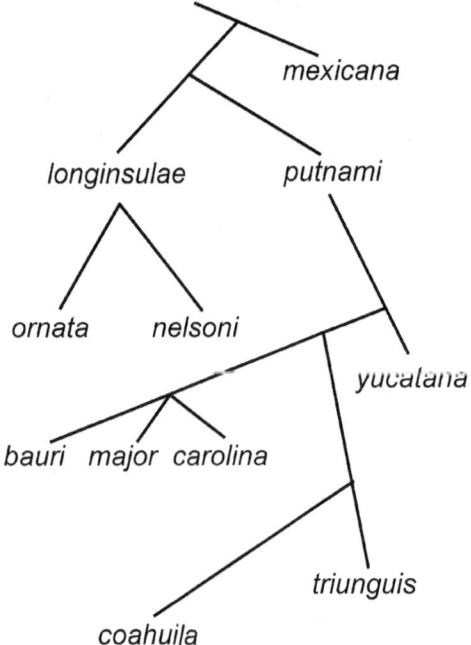

Phylogeny of *Terrapene*

118

The relationships of the box turtles have been rather confused; there seems to be wide variation in the characters of different populations and a considerable amount of overlap between supposed subspecies. In many areas the subspecies are difficult to identify. Many studies have come to quite contradictory conclusions of their relationships and full studies across the range of the genus have still not been completed. However, the available information does seem to be starting to point to one story of their origins, even though this does not fit easily with the conventional subspecies and species. Around 10 million years ago Carolina-like box turtles evolved in North America. The ancestral species seems to have given rise to the Mexican box turtle (currently *T. carolina mexicana*) and the rest of the living box turtles. The ancestor of these diverged into two significant fossil species: *T. longinsulae* and *T. putnami*. *T. longinsulae* is currently regarded as a subspecies of the ornate box turtle and seems to be ancestral to the ornate and to the spotted box turtle *T. nelsoni* of northern Mexico.

T. putnami is currently considered a subspecies of the Carolina box turtle but it may have been ancestral to the Carolina and the Coahuila box turtle *T. coahuila* of Coahuila state, Mexico. The first form to be isolated may have been *T. yucatana*, followed by a split between the *T. coahuila* and *T. triunguis* pair, and *T. bauri-major-carolina*. *T. yucatana* is the most southerly box turtle and the relationships of this group correspond to a northwards spread from the Yucatan peninsula (*T. yucatana*) to central Mexico-Texas (*T. coahuila* and *T. triunguis*) and east USA (*T. bauri-major-carolina*).

Within the eastern USA group there is considerable confusion, particularly as regards the identity of *T. carolina major*. Most *major* turtles are genetically identifiable as other subspecies and the range of *major* is best regarded as the area of interbreeding between the different subspecies. The area just to the north-west of the Florida peninsula (the 'Florida panhandle') is the main area of mixing, and seems to be where the interbreeding has been occurring for longest. Some turtles from this area seem to be more distinctive genetically and may be the closest there is to a true '*major*' population. These in turn could be living remnants of *T. putnami*. This was like *T. c. major*, but larger, exceeding 20 cm compared to *major*'s 17-20 cm and all other

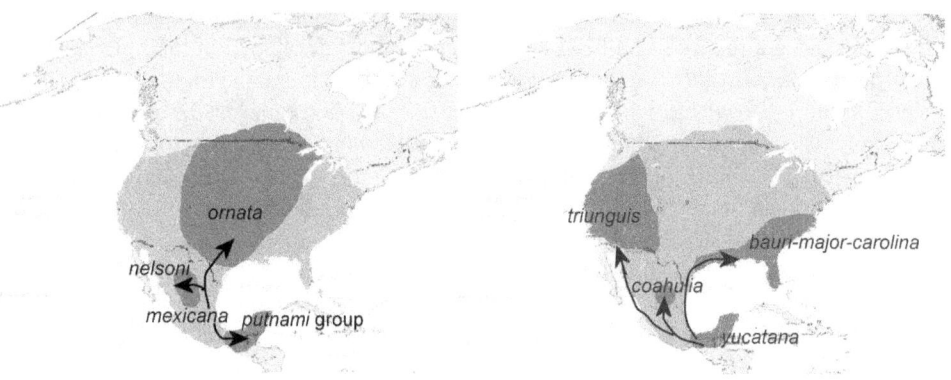

Evolution of *Terrapene* species; origins of the major groups on the left, diversification of the *putnami* group on the right.

subspecies to produce a smaller form, now called *major*. There is evidence of a decrease in the size of box turtles over the past 50,000 years, with average plastron measurements reducing from 10 cm 40,000 years ago to 9 cm 14,000 years ago and to the present day 7 cm.

In most box turtle species there is a tendency for the toes to become reduced through the loss of terminal bones of the outer toes. At an extreme this results in the complete loss of the outermost toe entirely, although in most 'three-toed' box turtles the fourth toe of the hind foot is present but reduced to a single bone; the three 'toes' really being claws. The ancestral box turtle probably had the most complete toes of all box turtles, with four claws on each foot and a fore-foot formula of 2-3-3-3-3 (number of phalangeal bones in each of the five toes going from the inner to the outermost) and a hind-foot formula of 2-3-3-3-2. The tendency towards toe reduction has occurred repeatedly in box turtles; in *T. mexicana*, in the *longinsulae* descendants and in the *T. putnami* descendants. In *T. mexicana* the fore-feet are slightly reduced (2-3-3-2-2) and the hind-feet more reduced, usually with three claws and a formula of 2-3-3-2-1 (some *mexicana* lose the outer toe completely). The descendants of *T. longinsulae* show strong reduction with the fore-feet reduced to 2-2-2-2-2 and hind-feet to 2-3-3-3-1 (in *T. ornata* and *T. nelsoni*). One exception is *T. n. klaueri* which has 2-3-3-3-3

The *T. putnami* group probably started with only slight reduction of the first toe (forefoot: 2-3-3-3-3, hindfoot: 2-3-3-3-2) and the living forms show independent reduction of the front toes in *T. coahuila*+*T. triunguis* and in *bauri*, and of the hind toes in *T. triunguis* and *T. bauri-major-carolina*. *T. yucatana* tends to have strong reduction of the outer hind-toe (2-3-3-3-1) but no fore-foot reduction. Both *T. coahuila* and *T. triunguis* have slight reduction (fore: 2-3-3-3-2), and some *T. triunguis* are more reduced (fore: 2-3-3-2-2, hind 2-3-3-3-1). In *T. bauri-major-carolina* there is usually some slight fore-foot reduction (2-3-3-3-2) and more in the hind feet (down to 2-3-3-2-1), although some *T. carolina* have full fore-foot development (2-3-3-3-3), and others are more reduced (2-3-3-2-2). *T. c. bauri* tends to have three claws on the hind feet, as is occasionally the case for *T. c. major* as well.

The postorbital bone and the bones of the cheek region (jugal and quadratojugal) of the skull are reduced in all box turtles but are most developed in the most aquatic forms: *T. coahuila* and *T. carolina major*. At the other extreme the highly terrestrial *T. ornata* and *T. nelsoni* have the most reduced postorbitals and have lost the quadratojugals entirely. The other species have narrow postorbitals and have the quadratojugal replaced by a piece of cartilage. This repeated modification of the feet in relation to the degree of terrestriality and reduction of skull bones associated with feeding specialisations make evolutionary reconstruction difficult when based on morphology alone.

The *Actinemys-Emys-Emydoidea* group evolved around 20 million years ago in North America. A recent study has suggested that DNA studies that show conflicts between mitochondrial and nuclear genes in the relationships between *Actinemys*, *Emys*, and *Emydoidea* may be due to hybridization between *Actinemys* and *Emydoidea*. Alternatively the same patterns could be the result of incomplete lineage sorting, where genes that are typical of one form have been inherited by another form from their common ancestor. Over time the frequency of these shared genes will decline, but at present a rather confusing pattern remains where nuclear genes say these are well supported genera but mitochondrial genes say otherwise. Even with genetic analysis the position of *Actinemys* is not clear. It has been proposed that this species, along with *Emydoidea*, should be placed in *Emys*, but this is still far from certain as *Actinemys* frequently groups with *Clemmys* rather than *Emys*. The most plausible explanation for the contradictions is that this group was originally restricted to North America (*Actinemys* and *Emydoidea*), colonising Eurasia about 16 million years ago via a trans-Beringian land bridge (giving rise to *Emys*). Further movement into Eurasia 12 million years ago may have resulted in hybridization between *Emydodiea* and *Emys*, resulting in nuclear genes showing a relationship between these two species and giving the conflicting results.

 Actinemys has one living species, the western pond turtle *A. marmorata*. This species is known from 5 million years ago from fossils described as '*Clemmys hesperia*' and also from an apparently different fossil species, *A. owyheensis*, from the same time. The living species is restricted to streams and marshes in western North America. It is one of only two freshwater turtle species found on the Pacific coast of the USA. Although nuclear DNA shows very little variaiton in this species, mitochondrial DNA shows four geographical groups: northern (northern California), San Joaquin Valley (southern Great Central Valley), Santa Barbara (Santa Barbara and Ventura counties) and southern (south of the Tehachapi Mountains and west of the Transverse Range south to Baja California, Mexico). The subspecies *A. marmorata* corresponds to the northern group and *A. pallida* to the three other groups, meaning that the currently recognised subspecies do not accurately reflect the diversity of this species. The genetic forms are divided by the Tehachapi mountains and the Transverse Range at the southern end of the Great Central Valley, as is found in other aquatic animals. The populations appear to have been separated about 4 million years ago when the Great Central Valley of California was flooded and formed a marine bay. There is also some evidence of more recent gene flow from the northern and San Joaquin Valley populations into the Central Coast Ranges after the isolation of the valley from the sea 2.5 million years ago, allowing movement of animals all around the valley. This species has also been introduced to New South Wales in Australia.

Actinemys

Emydoidea dates from 14 million years ago, with fossils of *E. hutchisoni* from Nebraska (this species appears to have become extinct 9 million years ago). This species was similar to the living Blanding's turtle *E. blandingii* but lacked mobility of the plastron found in the living species. This has been interpreted as indicating that this species was ancestral to *Emys, Emydoidea* and *Terrapene* but plastral mobility (kinesis) was probably independently evolved in *Emys* and *Terrapene.* Blanding's is known from around 5 million years ago. It is a largely carnivorous species always associated with water. It is mainly restricted to the north-central states of the USA, extending into southern Canada. There is also an isolated eastern population in the states of New York, Massachusetts, New Hampshire and Maine.

The ancestor of *Emys* had migrated to Europe by 20 million years ago by crossing the Bering Bridge between Alaska and Siberia. Shortly before 20 million years ago this was an area of dry land, with a cool but temperate climate. After 20 million years ago colder climates would have made any crossing impossible. The ancestral *Emys* is presumed to have moved westwards from Asia, into Europe. The oldest probable *Emys* fossils are from Kazakhstan 12 million years ago. In Europe the first true *Emys* fossils are *E. (orbicularis) wermuthi* from Poland around 4 million years ago. This was a relatively small, thin-shelled *Emys*, with a high shell and has been suggested to have been semi-terrestrial. This overlaps with *E. orbicularis antiqua* which occurred from 4 to 2.6 million years ago. This was slightly larger than the living pond turtles and was found from Poland to the North Caucasus. Pond turtle fossils from western Europe appear slightly later (within the last 800,000 years), with fossils from Italy, France and

England. By around 4 million years ago pond turtles were probably spread throughout Europe and North Africa. At about 5.5 million years ago the Mediterranean was isolated from the Atlantic and formed a large salt lake. This 'Messinian salinity crisis' would have isolated

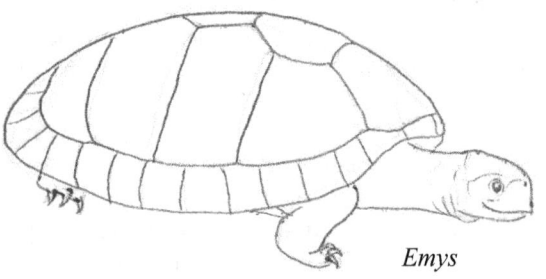

Emys

any turtle populations on the Mediterranean islands in the salt lake. This may be the origin of the Sicilian pond turtle *E. trinacris* which is superficially similar to Italian pond turtles but genetically distinctive, and the two populations seem not to interbreed in the small area of Italy where they come into contact. 5 million years ago the the Atlantic broke through the Straits of Gibraltar, flooding the Mediterranean basin in the catastrophic 'Zanclean Flood' that lasted 100 years.

As with so many European species, the origins of the pond turtle subspecies lie in the Ice Ages. In mainland Europe pond turtle ranges have fluctuated widely due to changing climatic conditions in the Ice Ages. As glaciers expanded, pond turtles retreated southwards into southern Europe. As the glaciers retreated the turtles moved back northwards again, only to retreat in the next cold phase. During the maximum extent of cold conditions in the last great Ice Age from 4 to 3.2 million years ago, pond turtles would have been forced into small areas of southern Europe that retained a relatively mild climate. From 4 million years ago to 3 million years ago they would have been restricted to refugia along the shores of the Mediterranean. When warmer conditions returned the turtles spread northwards, with fossils of pond turtles from the British isles from 120-110,000 years ago.

Further ice sheet expansion occurred again 20,000 years ago, forcing the pond turtles back into the southern areas. These refugia would have been in Spain-North Africa (giving the Ibero-Magrhebinian subspecies *E. orbiculalaris occidentalis-fritzjuergenobsti* group and *E. o. galloitalica*), south Sardinia, Italy (west coast – *E. o. galloitalica*; east coast – *E. o. hellenica*), Sicily (*E. trinacris*), the Balkans (west Greece – *E. o. hellenica*; east Greece - *E. o. orbicularis*), southern Turkey (*E. o. orbicularis*) and, further east, the southern shores of the Caspian Sea (*E. o. iberica*). Warmer conditions would have allowed recolonisation of their old range, initially this was slow due to physical barriers blocking some routes and initial conditions being too cold for the expansion of this Mediterranean species. The Balkans population seems to have spread northwards, *E. o. orbicularis* moving probably via the Danube river to the Baltic coasts of northern Germany and Poland. Some turtles from the northern Black Sea coasts also reached eastern Poland. Living populations from Germany and western Poland are genetically very similar and it seems only a small number of pond turtles colonised this area, probably as a result of flood waters washing a few most of the length of the Danube, resulting in very rapid range expansion northwards.

Further expansion westwards and northwards occurred more slowly. Denmark, Sweden and Britain were the latest parts of Europe to be colonised following the slow retreat of ice and cold conditions in Scandanavia 10,000 years ago. Jutland, the Danish islands and Sweden were a continuous area of dry land, forming the western shores of a vast freshwater lake, which is today the Baltic Sea. Pond turtles would have spread

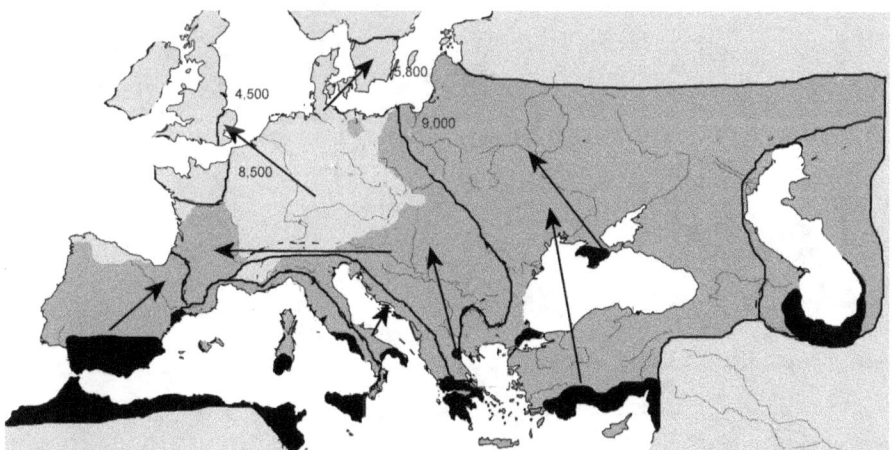

Expansion of *Emys orbicularis* in Europe over the past 10,000 years, showing the range of the main genetic groups and probable dispersal routes from Ice Age refugia (black) to the present range (grey) and beyond. Approximate dates of colonisation shown.

around the shores of this lake with little difficulty, being limited in their spread only by cool northern temperatures. At their greatest spread they reached 59° north (the latitude of Stockholm) around 5,800 years ago. By 4,500 years ago they had reached Britain over the dry land of the English Channel and North Sea exposed by the retreating ice (fossils from Norfolk). The most northerly populations were only temporary; between 5,000 and 2,000 years ago the climate cooled, the cool summer temperatures probably prevented successful reproduction and the northern populations declined. Although warmer conditions did recur subsequently, recolonisation was impossible as 8,500 years ago the North Sea flooded into the Baltic, disrupting the freshwater environment and further cooling local temperatures and leading to the final extinction of the Baltic pond turtles. The most recent Swedish remains are 5,500 years old. Similarly cold conditions eliminated the British pond turtles and flooding of the English Channel prevented later recolonisation.

Although most of the range expansion seems to have been through rivers and over dry land, some populations must have resulted from crossing stretches of open sea. Although pond turtles could have been present in North Africa from 5 million years ago when Spain and North Africa were in contact, the living populations of Spain and Morocco are genetically indistinguishable. This suggests that the North African *E. o. occidentalis* is descended from recent colonists of *E. o. fritzjuergenobsti* from Spain swimming across the Straits of Gibraltar. Although this seems highly unlikely there is currently no more plausible alternative explanation. Corsican and Sardinian pond turtles have been described as distinct subspecies (*E. o. lanzai* and *E. o. capolongoi* respectively) but are indistinguishable from *E. o. galloitalica* of western Italy and are thought to be recent introductions; however, these islands do have fossil pond turtle

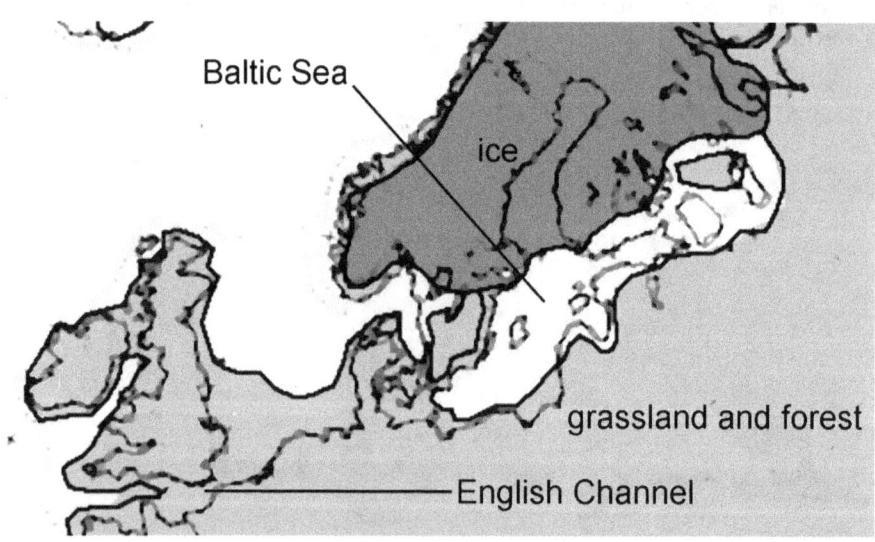

Northern Europe 10,000 years ago, showing the freshwater Baltic lake and dry English Channel.

remains from around a million years ago, suggesting an early natural colonisation followed by extinction and more recent introduction. It might be assumed that the natural colonists swam from Italy but as southern Sardinia would have been an ice-free refuge during the Ice Ages it is possible that the Corsican-Sardinian population was an ancient form like the Sicilian pond turtle. In the absence of remains that could provide DNA samples for comparison their origin will remain speculative.

As these various populations expanded following the retreat of the ice they came into contact in some areas, most notably around the Pyrenees mountains between France and Spain, along the French Mediterranean coast and in southern Italy. The Iberian *E. o. fritzjuergenobsti* was unable to cross the Pyrenees for the most part, but did manage to do so at the north-western edge where there is a gap between the mountains and the sea. In this area they came into contact with *E. o. orbicularis*. In the French Mediterranean *E. o. fritzjuergenobsti*, *E. o. galloitalica* and *E. o. orbicularis* met. In Italy *E. o. galloitalica* and *E. o. hellenica* came into contact. These hybrid zones led to an even local mixing of genetic forms except where *E. o. orbicularis* was present. In those places the *E. o. fritzjuergenobsti* and *E. o. galloitalica* populations contain mitochondrial genes of *E. o. orbicularis* but not the other way round. This may be because of a general geographical size difference in the European pond turtle; southern subspecies (and also the Sicilian pond turtle) are small and lighter than the northern populations (12-17cm compared to 16-23cm). They also lay two to three small clutches of eggs compared to the northern ones that lay one large clutch (23 eggs compared to 4-8). Where the northern *E. o. orbicularis* comes into contact with the southern subspecies, the large *E. o. orbicularis* females are sought out by the males as all male pond turtles prefer large females. Thus the flow of genes between the subspecies is only in one direction. Interbreeding has also occurred in the Calabria region of Italy where the distinct southern Italian genetic form of *E. o. galloitalica* mixes with *E. o. hellenica*. The Ligurian pond turtle *E. o. ingauna* occupies a small area of Italy, within the range of the *E. o. galloitalica* subspecies; at present it is not clear if this subspecies is valid.

Sicily is occupied by the isolated *E. trinacris*, which is genetically sufficiently distinct to be a separate species. This has also occupied the southern tip of Italy. This was the first pond turtle population to be isolated, and is thus the most distinctive genetically.

Pond turtles from the Balkans (*E. o. hellenica*) would have moved from an ancestral area around Croatia west via the Danube river, along a northwards route into central Europe, east towards Lithuania and the Caspian Sea, and west down into south-western France. Most of these northern and western European pond turtles have been considered to be the nominate subspecies *E. o. orbicularis*, which seems to be genetically identical to *E. o. hellenica*. This subspecies was isolated after *E. trinacris* and before all the others. *E. o. hellenica* would have been able to cross the Adriatic Sea around 18,000-14,000 years ago when low sea levels meant that this area of sea was much reduced in extent. By 6,000 years ago this would have been impossible. It has been suggested that the movement was from Italy to the Balkans but DNA data shows that the Italian *E. o. hellenica* are contained within the more diverse Croatian *E. o. hellenica*, suggesting that Croatian stock gave rise to the Italian.

125

From Turkey and the Levant the subspecies *E. o. eiselti* expanded north and east, occupying all of Turkey. Within Turkey another subspecies has been described – *E. o. luteofusca*. This is not distinguishable from *E. o. eiselti*. The Caspian refugium enabled expansion to Armenia, Iran and Turkmenistan for *E. o. persica*. This area also contains the subspecies *E. o. iberica* which is identical to *E. o. persica*. Nuclear DNA shows a relationship between *E. o. persica* and *E. o. orbicularis*, which indicates that the *E. o. orbicularis* moving east towards the Caspian Sea interbred with the *E. o. persica* in that area.

This is a highly aquatic species and its dispersal depends entirely on the availability of connections between water courses. It is often secretive and is associated with very clean water and is not tolerant of pollution; as a result it has declined in much of its range.

The slider and cooter group includes *Deirochelys* as the first offshoot, followed by *Chrysemys*, then *Pseudemys* and a group of *Trachemys, Malaclemys* and *Graptemys*. These are all typical highly aquatic terrapins and include some of the commonest turtle species. All are patterned and often have patches of bright colour. This is at least partly associated with mating as the intensity of colour pattern is often sexually dimorphic and males of all species have long claws which they tap against the head of the female or draw across her cheeks in courtship.

Deirochelys originated some 12 million years ago with the fossil species *D. carri* of Florida. The only living species is the chicken turtle *D. reticularia* from the southern and eastern coastal states of USA. This has distinct eastern (*D. r. reticularia*) and western (*D. r. miaria*) subspecies. In addition the Florida population is considered a third subspecies (*D. r. chrysea*); these differ most in colour pattern. The subspecies appear to have originated during the last Ice Age when cold, dry periods resulted in isolation of many rivers, alternating with periods of high sea-level when much of Florida was isolated as an island and coastal areas of the rivers were flooded with sea-water. The name 'chicken turtle' is a reference to the use of this species as an abundant food source in the past. It is rarely eaten by humans now, although juveniles are heavily predated by raccoons and alligators.

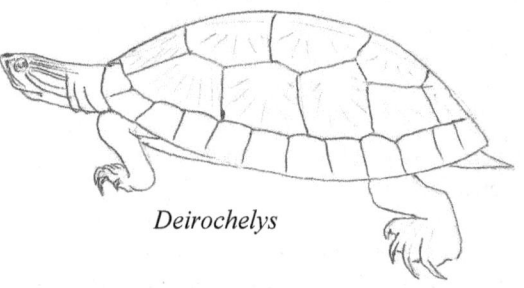

Deirochelys

Chrysemys seems to originate some 5 million years ago (the oldest reliable fossils being dated to somewhere between 10 and 5 million years ago). The living species are estimated to have diverged at around this time, resulting in living species having high levels of genetic divergence (2-2.5%). The oldest fossils attributed to *Chrysemys* are *C. antiqua* from South Dakota 34 million years ago but this is at least 20 million years older than the next fossils and this single fragmentary shell is probably a misidentification. Fossils of *C. timida* 5 million years ago may be a subspecies of the living *C. picta*. The painted terrapin *C. picta* is known from 8 million years ago

126

and more recently from throughout much of North America, except for the southern Mississippi drainage where it is replaced by the southern painted turtle *C. dorsalis*. Until recently it was thought that the only living species was the painted terrapin but that it included several subspecies. Mitochondrial DNA suggest that the painted and southern painted terrapins should be considered distinct species. Genetic studies suggest that *Chrsyemys*

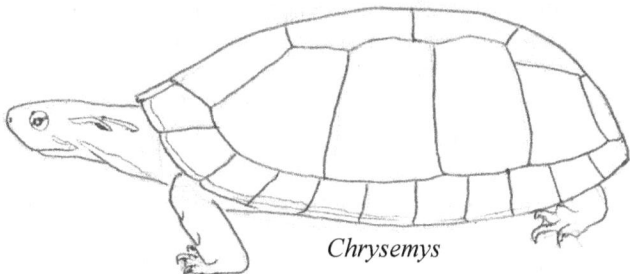

Chrysemys

species were widespread in North America 1 to 2 million years ago but died out at the end of the Ice Ages when extreme aridification over the central part of North America (the Great Plains and Rocky Mountains) 14,000 years ago would have eliminated many aquatic species. This left isolated populations in the southern Mississippi (giving rise to the southern painted turtle) and elsewhere. The latter subsequently re-invaded the ancestral range, giving rise to distinct subspecies in the east (*C. p. picta*), centre (*C. p. marginata*) and west (*C. p. bellii*). The painted turtle has been introduced to Germany, Spain, Indonesia and to California. This has been successful because it is a highly adaptable species, and naturally has the most widespread range of any North American turtle. In northern latitudes they are able to hibernate for prolonged periods, which they do underwater in waters as cold as 4°C.

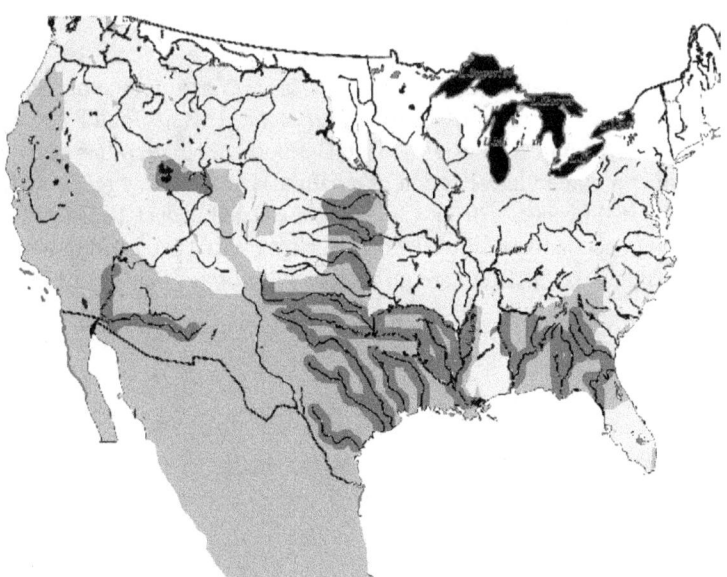

Location of probable river refugia in north America during the last Ice Age (white – ice; pale grey – cold or dry habitat; dark grey – refuges).

The relationships of the main genera of Emydidae (*Pseudemys, Trachemys, Malaclemys* and *Graptemys*) are not clear. Even within these genera there is much confusion. Mitochondrial and nuclear DNA studies often provide conflicting results with turtles; this is particularly apparent in the Emydidae. In particular, mitochondrial DNA tends to be very similar across different turtle species: *Graptemys* and *Pseudemys* species that are clearly distinct in their nuclear genes are almost identical in the mitochondria. Within the nuclear DNA itself, contradictory results can often be obtained by using different genes.

Pseudemys appears in the fossil record around 10 million years ago (*P. hilli* from Oklahoma, and the Floridan *P. caelata, P. inflata* and the living *P. nelsoni*). It now contains two species groups in the west and east. The western species are the Texas cooter *P. texana* and the Rio Grande cooter *P. gorzugi* from Mexico to southern Texas. The Eastern group comprises the river cooters and the red-bellied cooters. River cooters include the river cooter *P. concinna* (widespread in the USA) and the Peninsula cooter *P. peninsularis* (Florida). The red-bellied cooters include the Alabama red-bellied cooter *P. alabamensis* (Alabama and Mississippi), Florida red-bellied cooter *P. nelsoni* (Florida and Georgia, introduced into Texas) and the northern red-bellied cooter *P. rubriventris* (eastern USA). The relationships of the eastern forms are unclear; the Peninsula cooter is sometimes regarded as a subspecies of the coastal plain cooter *P. floridana* which is itself sometimes regarded as a separate species or as the coastal subspecies of the river cooter. Typical river cooters are found in the southern states of the USA, except in the coastal regions and in parts of Florida where they are replaced by the Suwannee cooter *P. c. suwanniensis*. This form has high-domed, thick shells in comparison to other river cooters, which may give them sufficient strength to withstand crushing from the American alligators that are common in these waters.

Within *Trachemys-Malaclemys-Graptemys* there seems to be a close relationship between *Malaclemys* and *Graptemys*. *Malaclemys* contains only one species, the diamondback terrapin *Malaclemys terrapin*. This is found throughout eastern USA and on Bermuda. The origin of the Bermuda population has been much disputed, being either a human introduction or a natural, but relatively recent colonist. The oldest remains of this species from the island are dated at AD 1427-1620. This overlaps with human occupation of the islands, which were first recorded in about AD 1500 but not settled until 1609. There are no records of terrapins being consumed and

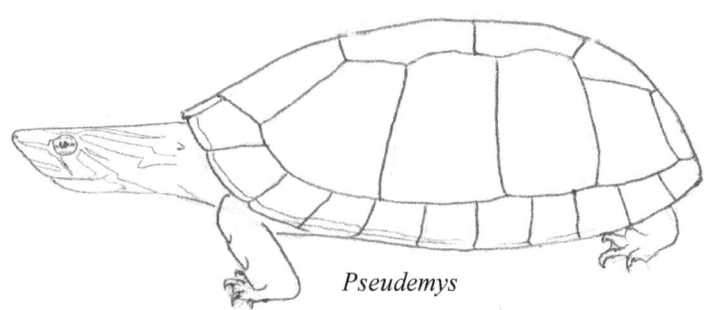

Pseudemys

transported by European settlers until the 1800s. Genetic data suggest that Bermudian terrapins originated in coastal Carolina. This fits with the Gulf Stream washing terrapins from that region of North America to Bermuda. There is very little genetic difference between the mainland and island populations, which supports the view that they are recent in origin. If they are natural colonists this would probably have occurred within the last 3,000 years when the mangrove habitats they occupy developed on the island. They are well adapted to be natural colonists of the islands, being found in brackish habitats: having well developed lachrymal glands that can excrete salt they are able to tolerate high salinities.

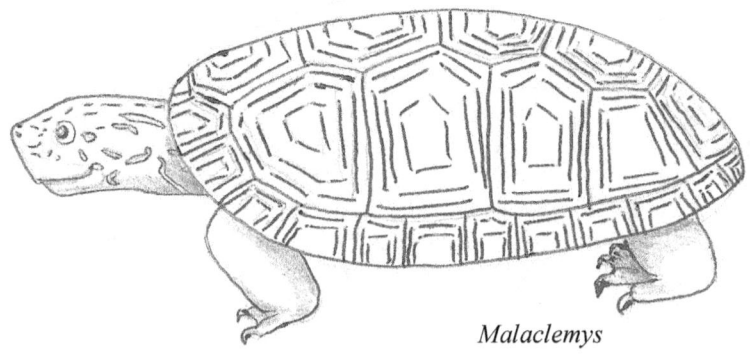

Malaclemys

Mainland diamondback populations date back at least 2 million years (fossils from South Carolina) and are divided into several subspecies. The primary division is between the western (the Texas diamondback *M. t. littoralis* in Texas) and the eastern populations. The main eastern population comprises the Northern diamondback *M. t. terrapin* in most of the range and the Carolina diamondback *M. t. centrata* in the southeast, except for Florida. In Florida there is a complex of three subspecies: the ornate diamondback *M. t. macrospilota*, mangrove diamondback *M. t. rhizophorarum* and Eastern Florida diamondback *M. t. tequesta*. In addition the Mississippi drainage is occupied by the Mississippi diamondback *M. t. pileata*.

The relationships between *Graptemys* and *Trachemys* species are far from clear. In *Graptemys* the northern or common map turtle *G. geographica* is the first species to diverge, followed by two major groups: the Alabama map turtle group, or 'megacephalic' group (Alabama map turtle *G. pulchra*, Escambia *G.. ernsti*, Pascagoula *G. gibbonsi*, Barbour's *G. barbouri*, Pearl river *G. pearlensis*); and the sawbacked terrapins (ringed *G. oculifera*, yellow-blotched *G. flavimaculata*, black-knobbed *G. nigrinoda*, Ouachita *G.. ouachitensis,* false map turtle *G. pseudogeographica*, Cagle's *G.. caglei* and Texas *G. versa*). The Alabama map turtle group have been called the 'megacephalic' group because the females have very large heads, with remarkably wide jaws. The females feed almost exclusively on molluscs (pond snails and freshwater clams) species; males will also feed on molluscs but are not strict molluscivores and do not have the same skull features. Females are also considerably larger than males (to 33 cm compared to 13 cm). This group originated some 3 million years ago (as indicated by fossils ascribed to *G. 'barbouri'*).

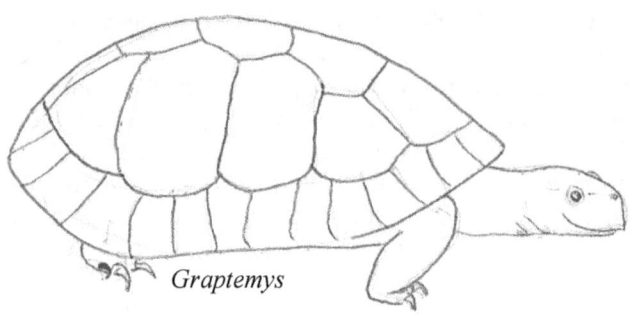

Graptemys

All members of this group are restricted to different river drainages by rising sea levels at the end of the Wisconsinian Ice Age 20,000 years ago. This would have flooded coastal marshes which had previously allowed terrapins to migrate between river drainage systems. The eastern-most species (which are also the biggest headed) are Barbour's in the Apalachicola river system and the Escambia map turtle in the Escambia river, both extend into the intervening Choctawhatchee. In the centre of the range is the Alabama map turtle; to the west the Pascagoula and the Pearl, each in their separate river systems. In addition to the living species there is a fossil species from the east of the range; *G. kerneri* is closely related to Barbour's map turtle and is dated from 15,000 years ago from the Big Bend region of Florida. The rivers that were occupied by this species are isolated from the rest of the range of the genus but the Suwannee river would probably have been in contact with the Apalachicola 20,000 years ago when sea levels were lower. The flooding of this confluence of rivers isolated Barbour's and *G. kerneri*. Subsequent climate change may have led to a deterioration of conditions in the Suwannee and the eventual extinction of *G. kerneri*. There is a fossil species that is superficially similar but is probably not related: *Pseudograptemys inornata* is known from South Dakota 35 million years ago. It is clearly an emydid turtle but it is impossible to determine its relationship to other emydids.

Within the sawbacks there are two further subgroups: the *oculifera* group (*G. oculifera, G. flavimaculata* and *G. nigrinoda*) and the *pseudogeographica* group (*G. ouachitensis, G. pseudogeographica, G. caglei* and *G. versa*). As with the Alabama map turtle group the *oculifera* group is isolated in different rivers of Alabama, Louisiana and Mississippi: the black-knobbed (*G. nigrinoda*) in the Alabama, the yellow-blotched in the Pascagoula (*G. flavimaculata*) and the ringed (*G. oculifera*) in the Pearl. The false map turtle group species are all very closely related and difficult to separate. There seems to be a discordance between current subspecies and genetic groupings: *G. ouachitensis* is ancestral to the other *pseudogeographica* taxa but is isolated from its supposed subspecies *G. o. sabinensis* which groups with the Texan species of Cagle's (*G. caglei*) and Texas (*G. versa*) map turtles. In the black-knobbed map turtle northern and southern populations are regarded as distinct subspecies found in different river drainages: *G. n. nigrinoda* in the Alabama and Tombigbee and *G. n. delticola* in the Mobile Bay drainage. The Ouachita map turtle has the nominate subspecies in most of its range with the Sabine map turtle *G. o. sabinensis* in Louisiana and Texas. The false map turtle has a distinct subspecies, the Mississippi map turtle *G. p. kohnii* in the wider Mississippi drainage.

130

Trachemys are the well-known sliders of North America, including the red-eared slider *T. scripta elegans*, the most widely introduced and traded turtle species. The earliest fossils include the red-eared slider from Florida and the fossil species *T. inflata* from Nebraska, both from around 3 million years ago. Mitochondrial DNA studies of *Trachemys* give a split

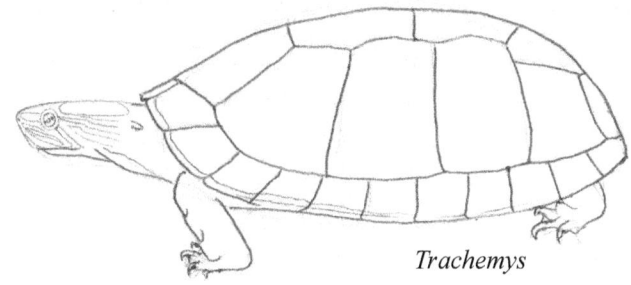

Trachemys

between North American and Central-South American species. Within North America the common slider *T. scripta* and the big bend slider *T. gaigeae* of Mexico and Texas are sister species. The big bend slider has been thought to be close to, or possibly synonymous with, South American or Mexican species but DNA suggests that it is a valid species close to *T. scripta*. It has been divided into northern (the big bend slider *T. g. gaigeae* from the Rio Grande and Rio Conchos of Texas, New Mexico and Mexico) and southern (Nazas slider *T. g. hartwegi* from the Rio Nazas in Mexico) subspecies. Within the common slider there is confusion between the red-eared slider *T. s. elegans*, the yellow-bellied slider *T. s. scripta* and the Cumberland slider *T. s. troostii*.

In the central-southern species three groups are identifiable: Mexico to Venezuela, Brazil, and the West Indies. Brazil has two species: D'Orbingy's slider *T. dorbigni* in the south, extending into northern Argentina, and the Maranhao slider *T. adiutrix* in the north, in Maranhao state. D'Orbigny's is a largely brown species in the south and west part of its range (*T. d. dorbigni*) or green in the Brazilian part of the range (*T. d. brasiliensis*). The Maranhao slider is highly restricted, living in lakes in a sandy dune area. Fossils possibly identifiable as D'Orbigny's slider have been found in central Brazil from 120,000 years ago.

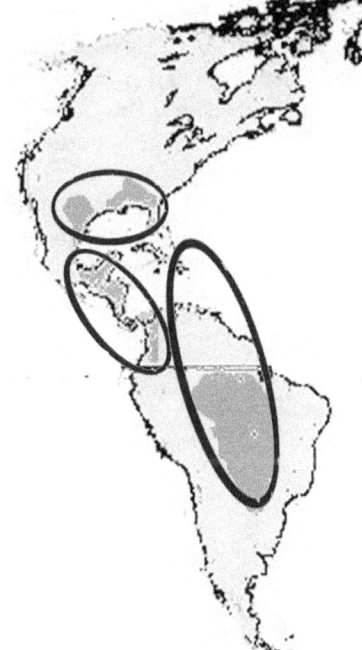

Distribution of the *Trachemys* groups.

The West Indies group comprises the Cuban slider *T. decussata* and a grouping of the Hispaniolan slider *T. decorata,* Central Antillean slider *T. stejnegeri* (Bahamas, Hispaniola and Puerto Rico) and the Jamaican slider *T. terrapen*. The Jamaican slider is present in Jamaica and on Cat Island in the Bahamas where it may be a prehistoric introduction. The Cuban slider is found in Cuba (the nominate subspecies) and in the

Cayman islands and parts of Cuba (*T. d. angusta*). This is the most unusual of the sliders, lacking the red or yellow head stripe found in most *Trachemys*, being covered in white spots instead. The Central Antillean slider includes three subspecies, *T. s. stejnegeri* from Puerto Rico, the Inagua slider *T. s. malonei* of Inagua Island in the Bahamas and the Dominican slider *T. s. vicina* from the island of Hispaniola. DNA suggests that Cuba was the first island to be occupied but the currents in the Caribbean make it probable that the first population was on Puerto Rico and that South American sliders were carried westwards along the island chain.

Central American sliders include the Nicaraguan slider *T. emolli* (Costa Rica and Nicaragua), followed by the Mexican Yaqui slider *T. yaquia*, Cuatro Ciénegas slider *T. taylori*, Baja California slider *T. nebulosa* and ornate slider *T. ornata* (western costal Mexico to Guatemala), and finally the Meso-American slider *T. venusta* and the Colombian-Venezuelan Colombian slider *T. callirostris*. The Baja Californian slider is found in four isolated areas in the arid west of Mexico. These areas support two distinct subspecies: the Baja California slider *T. n. nebulosa* from Baja California and the Fuerte Slider *T. n. hiltoni* from the Rio Fuerte in Sinaloa and Sonora. The Meso-American slider comprises the nominate subspecies from Belize to Mexico, the Huastecan slider *T. v. cataspila* from Mexico, Gray's slider *grayi* from Guatemala and south Mexico, Yucatan slider *T. v. iversoni* from the Yucatan of Mexico, Panamanian slider *T. v. panamensis* of Panama and Uhrig's slider *T. v. uhrigi* of Nicaragua and Panama. The Colombian slider is divided into the Colombian subspecies *T. c. callirostris* and the Venezuelan *T. c. chiochiririche*.

All of these diverse slider forms represent geographically isolated variants of a very successful animal. The distribution of the species reflects isolation of major areas in times of marine incursion into rivers and estuaries, with the subspecies evolving within river systems.

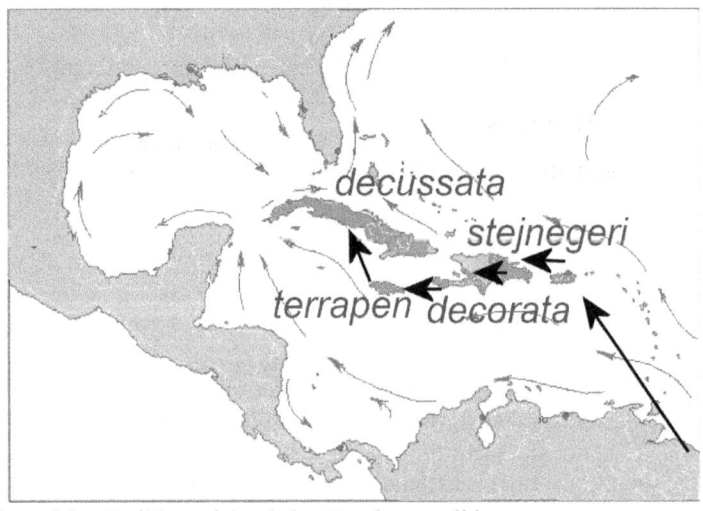

Colonisation of the Caribbean islands by *Trachemys* sliders.

References

Aresco, J.M. & J.L. Dobie. 2000. Variation in shell arching and sexual size dimorphism of river cooters, *Pseudemys concinna*, from two river systems in Alabama. *Journ. Herpetol.* **34**: 313-317

Ehret, D.J. & J.F. Bourque. 2011. An extinct map turtle Graptemys (Testudines, Emydidae) from the late Pleistocene of Florida. *Journ. Vert. Palaeont.* **31**: 575-587

Ennen, J.R., J.E. Lovich, B.R. Kreiser, W.W. Selman, II, & C.P. Qualls. 2010. Genetic and morphological variation between populations of the Pascagoula Map Turtle (*Graptemys gibbonsi*) in the Pearl and Pascagoula Rivers with description of a new species. *Chel. Conser. Biol.* **9**: 98–113

Feldman, C.R. & J.F. Parham. 2002. Molecular Phylogenetics of Emydine Turtles: Taxonomic Revision and the Evolution of Shell Kinesis. *Mol. Phyl. Evol.* **22**: 388-398

Fritz, U., A. Cadi, M. Cheylan, C. Coïc, M. Détaint, A. Olivier, E. Rosecchi, D. Guicking, P. Lenk, U. Joger, & M. Wink. 2005. Distribution of mtDNA haplotypes (cyt *b*) of *Emys orbicularis* in France and implications for postglacial recolonization. *Amphibia-Reptilia* **26**: 231-238

Fritz, U., T. Fattizzo, D. Guicking, S. Tripepi, M.G., Pennisi, P. Lenk, U. Joger & M. Wink. 2005. A new cryptic species of pond turtle from southern Italy, the hottest spot in the range of the genus *Emys* (Reptilia, Testudines, Emydidae). *Zool. Script.* **34**: 351–371

Holman, J.A. 2002. Additional Specimens of the Miocene Turtle *Emydoidea hutchisoni* Holman 1995 - New Temporal Occurrences, Taxonomic Characters, and Phylogenetic Inferences. *Journ. Herpetol.* **36**: 436-446

Holman, J.A. & U. Fritz. 2005. The Box Turtle Genus *Terrapene* (testudines: Emydidae) in the Miocene of the USA. *Herpetol. Journ.* **15**: 81-90

Holman, J.A. & U. Fritz. 2001. A new emydine species from the Medial Miocene (Barstovian) of Nebraska, USA with a new generic arrangement for the species of *Clemmys* sensu McDowell (1964) (Reptilia, Emydidae). *Zool. Abh. Staat. Mus. Tier. Dresen* **51**: 331-354

Jackson, J.T., D.E. Starkeym R.W. Guthrie & M.R.J. Forstner. 2008. Mitochondrial DNA phylogeny of extant species of the genus *Trachemys* with resulting taxonomic implications. *Chel. Cons. Biol.* **7**: 131-135

Lapparent de Broin, F. de. 2001. The European turtle fauna from the Triassic to the Present. *Dumerilia* **4**(3): 155–217

Lenk, P., U. Fritz, U. Joger & M. Wink. 1999. Mitochondrial phylogeography of the European pond turtle, *Emys orbicularis* (Linnaeus 1758). *Mol. Ecol.* **8**: 1911–1922

Parham, J.F., M.E. Outerbridge, B.L. Stuart, D.B. Wingate, H. Erlenkeuser & T.J. Papenfuss. 2008. Introduced delicacy or native species? A natural origin of Bermudian terrapins supported by fossil and genetic data. *Biol. Lett.* **4**: 216–219

Parham, J.F., W.B. Simison, K.H. Kozak, C.R. Feldman & H. Shi 2001. A reassessment of two species using mitochondrial DNA, allozyme electrophoresis and known-locality specimens. *Anim. Conserv.* **4**: 357-367

Pedall, I., U. Fritz, H. Stuckas, A. Valdeon & M. Wink. 2010. Gene flow across secondary contact zones of the *Emys orbicularis* complex in the Western Mediterranean and evidence for extinction and re-introduction of pond turtles on Corsica and Sardinia (Testudines: Emydidae). *J Zool Syst Evol Res*

Shaffer H.B., P. Meylan & M.L. McKnight. 1997. Tests of turtle phylogeny: Molecular, morphological, and paleontological approaches. *Syst. Biol.* **46**: 235-268

Sommer, R.S., C. Lindqvist, A. Persson, H. Bringsoe, A.G.J. Rhodin, N. Schneeweiss, P. Široky, L. Bachmann & U. Fritz. 2009. Unexpected early extinction of the European pond turtle (*Emys orbicularis*) in Sweden and climatic impact on its Holocene range. *Mol. Ecol.* **18**: 1252-1262

Sommer, R.S., A. Persson, N. Wieseke & U. Fritz. 2007. Holocene recolonization and extinction of the pond turtle, *Emys orbicularis* (L. 1758), in Europe. *Quatern. Sci. Rev.* **26**: 3099-3107

Sommer, R.S., U. Fritz, H. Seppa, J. Ekstrom, A. Persson & R. Liljegren. 2011. When the pond turtle followed the reindeer: effect of the last extreme global warming event on the timing of faunal change in Northern Europe. *Global Change Biology* **17**(6): 2049-2053

Spinks, P.Q. & H.B. Shaffer. 2005. Range-wide molecular analysis of the western pond turtle (*Emys marmorata*): cryptic variation, isolation by distance, and their conservation implications. *Mol. Ecol.* **14**: 2047–2064

Spinks, P.Q. & H.B. Shaffer. 2009. Conflicting Mitochondrial and Nuclear Phylogenies for the Widely Disjunct *Emys* (Testudines: Emydidae) Species Complex, and What They Tell Us about Biogeography and Hybridization. *Syst. Biol.* **58**: 1-20

Spinks, P.Q., R.C. Thomson & H.B. Shaffer. 2010. Nuclear gene phylogeography reveals the historical legacy of an ancient inland sea on lineages of the western pond turtle, *Emys marmorata* in California. *Mol. Ecol.* **19**: 542-556

Starkey, D.E., H.B. Shafferm R.L. Burke, M.R.J. Forstner, J.B. Iverson, F.J. Janzen, A.G.J. Rhodin & G.R. Ultsch. 2003. Molecular systematics, phylogeography, and the effects of Pleistocene glaciation in the painted turtle (*Chrysemys picta*) complex. *Evolution* **57** (1):119-128

Thomson, R.C. & H.B. Shaffer. 2010. Sparse Supermatrices for Phylogenetic Inference: Taxonomy, Alignment, Rogue Taxa, and the Phylogeny of Living Turtles. *Syst. Biol.* **59**(1): 42–58

Wiens, J.J., C.A. Kuczynski & P.R. Stephens. 2010. Discordant mitochondrial and nuclear gene phylogenies in emydid turtles: implications for speciation and conservation. *Biol. Journ.Linn. Soc.* **99**: 445-461

12. River turtles – Geoemydidae

Geoemydidae (sometimes known as Bataguridae) is the largest turtle family. It contains *Rhinoclemmys, Geoemyda, Siebenrockiella, Ortilia, Malayemys, Geoclemys, Morenia, Pangashura, Hardella, Batagur, Melanochelys, Vijayachelys, Sacalia, Leucocephalon, Notochelys, Cyclemys, Heosemys Cuora* and *Mauremys*. They include terrestrial, semi-aquatic and aquatic turtles from the Americas, Asia, Europe and north Africa.

Mitochondrial and nuclear DNA identify two groupings within the family: old-world Geoemydidae, and the South American geoemydid genus *Rhinoclemmys*. Fossils attributable to this family date back as far as 55 million years ago, however, most of these fossils are poorly preserved and their identity must be regarded as uncertain. The ptychogastrid *Echatemys* from North America (55 million years ago) has been though to be ancestral to *Rhinoclemmys* on the basis of having musk ducts enclosed in the peripheral plates of the shell. This is characteristic of several, but not all, living geoemydids but is a primitive character of testudinoids, and takes a particularly good fossil to be clearly recognisable (such as *Palaeoemys testudiniformis* and *P. hessiaca*). *Echatemys* species were found throughout North America (from Wyoming and Ellesmere Island) by 50 million years ago. This genus has also been reported from Asia but these species (*E. orlovi, E. zaisanensis, E. chingaliensis* and *E. borisovi*) are almost certainly misidentified. *Echatemys* thrived in the extensive lakes and swamps that spread across North America during the warm, humid conditions of 50 million years ago. These gradually changed to a cooler, dryer climate and this combined with geological movement to result in the drying of most of the swamps. *Echatemys* declined and disappeared about 40 million years ago. The same fate befell the contemporaneous (from 49 to 46 million years ago) *Bridgeremys*, another possible early geoemydid. *Echatemys* has also been suggested to be ancestral to the Emydidae.

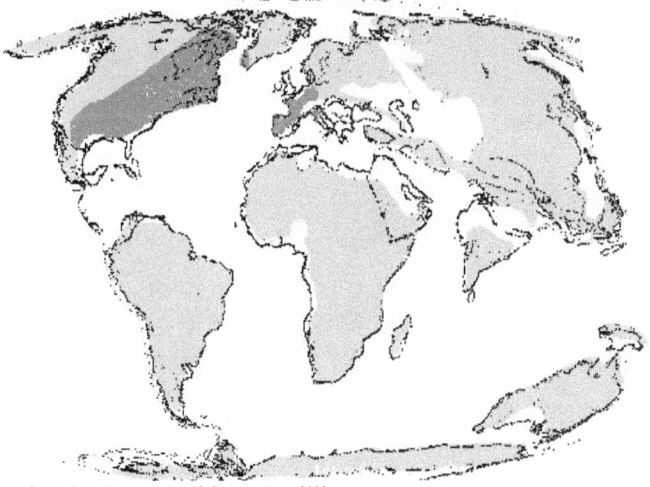

Distribution of early Geoemydidae 55 million years ago.

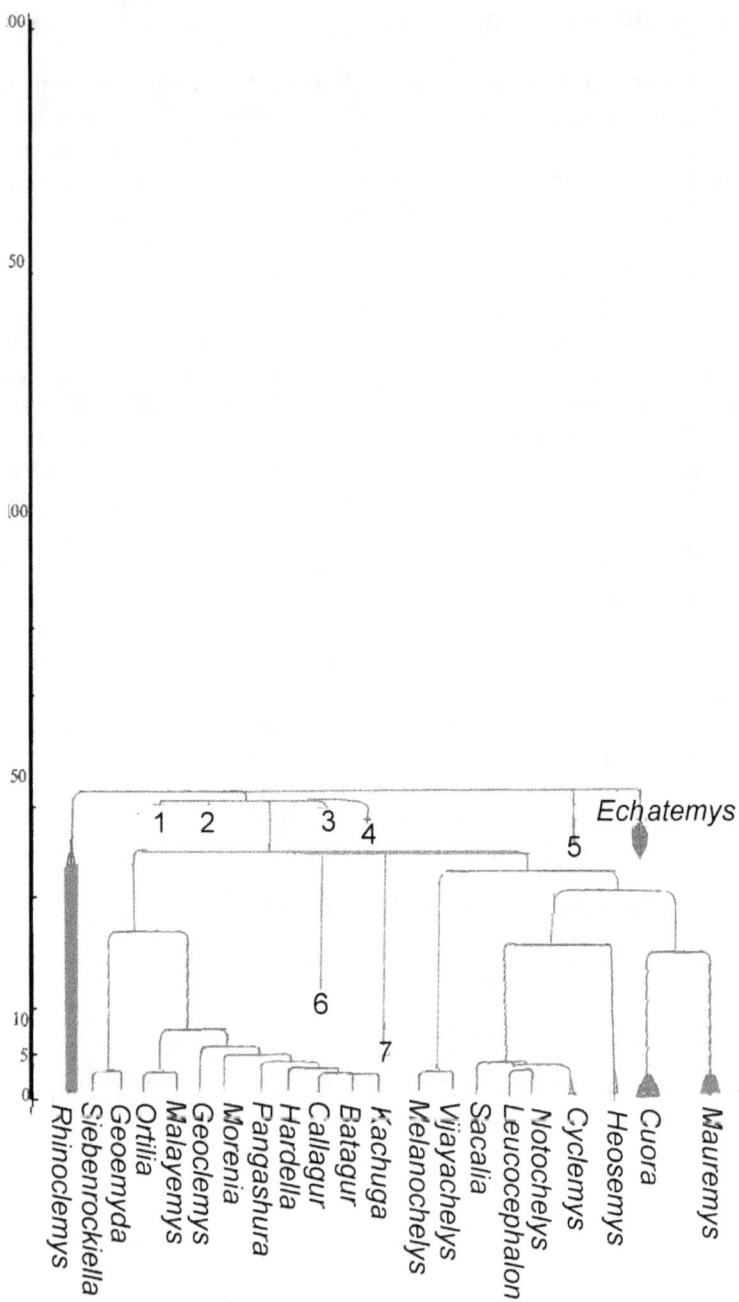

Geoemydidae phylogeny
1 - *Palaeoemys*; 2 - *Palaeochelys*; 3 - *Cuvierichelys*; 4 - *Grayemys*; 5 - *Bridgeremys*;
6 - *Sakya*; 7 - *Epiemys*

Geoemydidae seem to have arisen in the northern continents some 55 million years ago; the first definite geoemydids are from around 53 million years ago from France in the genera *Palaeoemys* and *Palaeochelys*. A more recent genus, *Borkenia* (45 million years ago from Germany) may be either a close relative of *Palaeoemys* or a synonym. The living American genus *Rhinoclemmys* appears slightly later in Utah 40 million years ago.

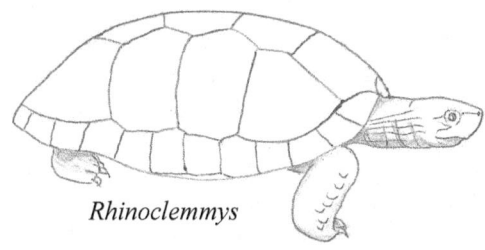

Rhinoclemmys

Rhinoclemmys are small to medium-sized turtles of forest streams. Along with the *Trachemys* sliders, they are the only cryptodire turtles to colonise South America, although they only just penetrate the continent. They probably originated in North America but were able to spread into Central America during the cooling at 50 million years ago. There was an early division between the closely related largely terrestrial species pair of the painted wood turtle *R. pulcherrima* and the brown wood turtle *R. annulata*, and the rest of the genus. The painted and brown wood turtles are found from the Pacific coastal regions of Mexico to Nicaragua and from Nicaragua to Ecuador respectively. These two species may have been isolated by the geological upheaval that formed the 'Nuclear Highlands' of southern Mexico (Chiapas), Guatemala and Honduras 15 million years ago, separating the north and south populations.

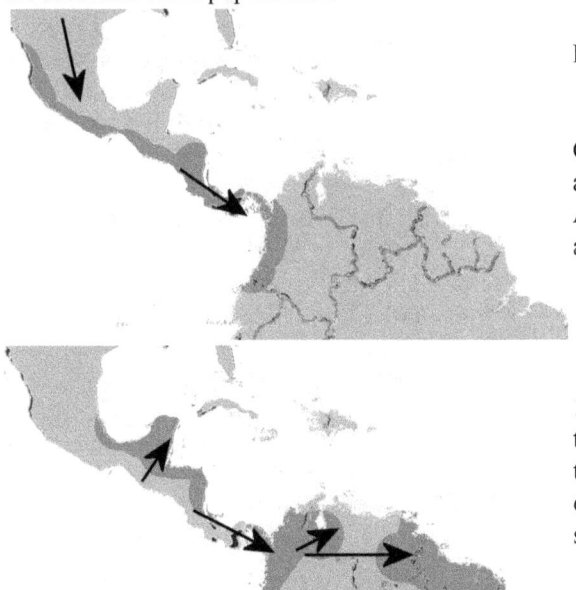

Distribution of *Rhinoclemmys*

1) Geoemydids move into Central America, spreading along the west coast to South America giving rise to painted and brown wood turtles.

2) *Rhinoclemmys* cross to the east coast and repeat the southwards movement, extending into north-eastern south America.

Of the remainder of the species the earliest of the living species was a small widespread coastal species of Mexico and possibly Guatemala, the Mexican spotted wood turtle *R. rubida*. This was followed by the large-nosed wood turtle *R. nasuta* of the Pacific coast of Colombia and Ecuador. This is a species of fast flowing rivers, unusual in turtles in that it lays its eggs without a nest and covers them with leaves. These species may have been isolated from the rest of the genus by the uplift of the eastern Sierra Madre mountains of Mexico 20 million years ago, separating the Mexican spotted and the large-nosed wood turtles on the Pacific lowlands from the populations on the Atlantic lowlands. The large-nosed wood turtle may have originated further north, and colonised its present range through island hopping. It now lives mainly in the Choco region, which was a series of volcanic islands off the coast of Central America about 15 million years ago. If the ancestor of the large-nosed wood turtle was a Central American species that was swept onto these volcanic islands it could have been transported by geological movement to the northern tip of South America as the Choco islands became consolidated into that continent 3 million years ago.

The Atlantic species comprise the furrowed wood turtle *R. areolata* (from the Yucatan peninsula) and several species found from Costa Rica southwards. Relatives of these species moved south towards South America as the Panama isthmus closed around 3 million years ago, giving rise to Panamanian and truly South American groups; these would have been isolated from the furrowed wood turtle by the formation of the Nuclear Highlands 15 million years ago. The Panamanian populations divided into the two largest *Rhinoclemmys* species: the Colombian wood turtle *R. melanosterna* (Ecuador to Panama) and the black wood turtle *R. funerea* (Costa Rica to Panama). The black wood turtle is the largest of all and is also unusual in geoemdyines in being completely herbivorous. The South American population diverged into the western Maracaibo wood turtle *R. diademata* (Colombia and Venezuela) and the eastern spot-legged turtle *R. punctularia* (Brazil to Venezuela and Trinidad). In the past this latter group may have been even more widespread as there are *Rhinoclemmys* fossils from Brazil further south than at present.

The painted wood turtle includes distinct subspecies in parts of Mexico (Guerrero wood turtle *R. p. pulcherrima*), Nicaragua to Mexico (incised wood turtle *R. p. incisa*), Costa Rica and Nicaragua (Central American wood turtle *R. p. manni*) and western Mexico (western Mexico wood turtle *R. p. rogerbarbouri*). Of these the Nicaraguan subspecies are the most closely related. The spot-legged turtle has two subspecies – the eastern spot–legged turtle *R. p. punctularia* in most of the range and the Upper Orinoco spot-legged turtle in the Orinoco river in Venezuela. The Mexican spotted wood turtle has the Oaxaca wood turtle *R. r. rubida* in Chiapas and Oaxaca states of Mexico, and the Colima wood turtle *perixantha* of Colima, Jalisco and Michoacán states. *Rhinoclemmys* are mainly aquatic turtles although the brown and painted wood turtles are largely terrestrial. All species have bright colour markings on the head at least; in the painted wood turtle distinctive patterns are also found on the shell.

Unlike most geoemydids the American *Rhinoclemmys* retains the primitive character of the absence of distinctive keels on the carapace. These keels seem to have arisen in the European geoemydids, being seen first in *Palaeoemys* from France 53

138

million years ago. This is followed by another French species, *Cuvierichelys parisensis* from 45 million years ago, and the Kazakhstani *Grayemys* from 40 million years ago. From 95 million years ago Europe and Asia were separated by the Obik (or Ural) Sea lying to the west of what are today the Ural mountains in Russia. This was narrowing progressively; reduced sea levels exposed temporary land bridges between Europe and Asia from 65 million years ago and by 40 million years ago the southern part was only a relatively narrow channel, the Turgai Straits. Presumably the ancestor of *Grayemys* crossed these straits to give rise to the Asian Geoemydidae 40 million years ago. 10 million years later the Obik Sea closed, resulting in the present day continuous Eurasian landmass, facilitating the expansion of the geoemydids to their modern range from the eastern edge of Europe to Japan. Fossils from this time include the eastern European to west Asian *Sakya* (from within the last 11 million years) and the Asian *Epiemys* (5 million years ago). The old-world geoemydid group is highly diverse in species and habits. Within this diversity there are two main groups: the '*Batagur* group' and the '*Cuora/Mauremys* group'. At the base of these lie the genera *Siebenrockiella* and *Geoemyda*. These are very hard to place and may group with either of the main groups. Both groups may have originated as large river turtles in the complex of rivers around the present day Ganges and Brahmaputra which formed when the Indian and Asian continental plates collided 20 million years ago. The *Batagur* group retain much of the ancestral habits and range but the *Cuora/Mauremys* group spread eastwards over the past 2.5 million years, adapting to life as semi-aquatic forest turtles.

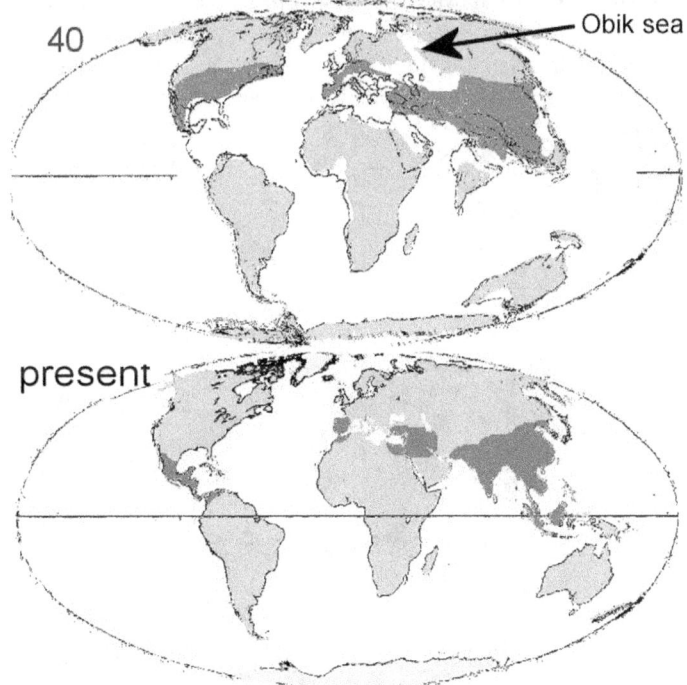

Distribution of Geoemydidae from 40 million years ago

The Philippine forest turtle *Siebenrockiella leytensis* is very poorly known. It is restricted to the Philippine island of Palawan. It was originally placed in *Heosemys* but DNA indicate that this is a sister species to the South-east Asian 'black marsh turtle' or 'smiling terrapin' *S. crassicollis*. These two

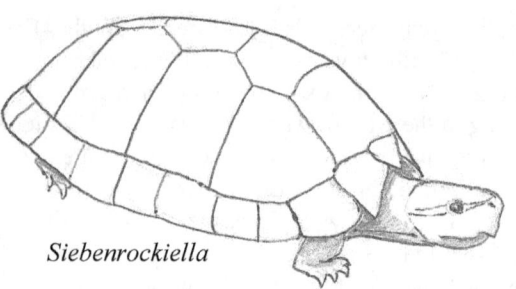

Siebenrockiella

taxa have distinctive mushroom- or 'ginkgo leaf'-shaped vertebral scutes. Along with some species of *Cuora*, *Terrapene*, *Heosemys* and *Hieremys* the Philippine forest turtle has an extreme degree of temporal emargination, lacking a temporal arch in the skull. Both are restricted to localised forest areas under threat from deforestation. The Philippine forest turtle and the smiling terrapin both occur in the islands of south-east Asia. The Philippine forest turtle is known only from the island of Palawan and nearby islands which would have formed a single large island from 5 to 2 million years ago when sea levels were lower than they are today. The smiling turtle is found in Malaysia, Thailand and Vietnam, and on Sumatra and Java. These islands of the Sunda Shelf (Borneo, Sumatra etc) would have been united with the large Palawan island, whereas the other Philippine islands did not share a land connection at the time. Several other animal and plant species show a similar pattern, with a closer relationship between Palawan and Borneo than between Palawan and other Philippine islands. Very little is known of the biology of the Philippine forest turtle. The smiling terrapin is small (20 cm), and dark coloured with a large, pale head. It is found in slow moving water bodies where it walks along the bottom. It is omnivorous and adaptable. Its strong-

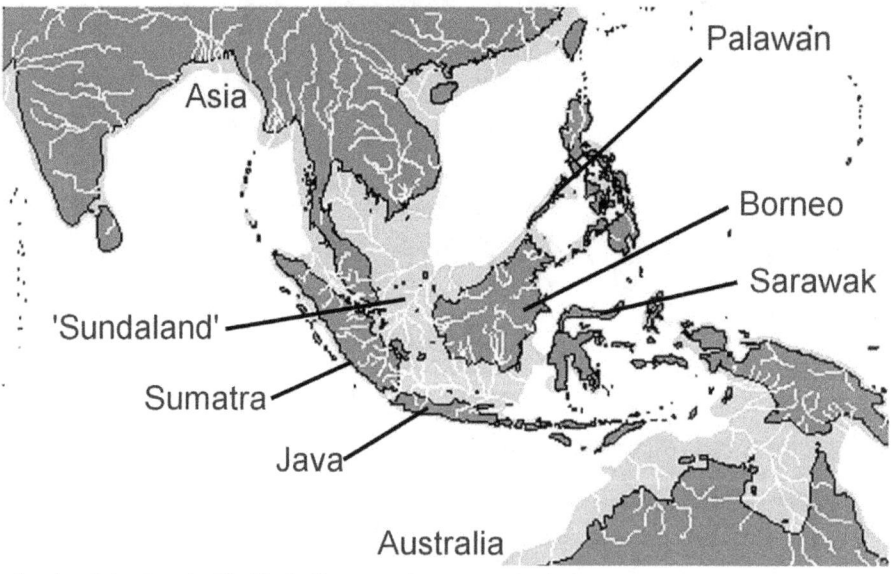

Islands of the Sunda Shelf, shallow marine areas shaded.

smelling defensive musk acts as an effective protection and they are rarely eaten by humans.

Geoemyda comprises two very distinctive endangered species: the Ryukyu black-breasted leaf turtle *G. japonica* of the Ryukyu archipelago (Japan) and the black-breasted leaf turtle *G. spengleri* of southern China and Vietnam. The black-breasted leaf turtle is a small (13 cm) species with a distinctively serrated margin to the shell and a prominent keel. The Japanese species is less serrated but more brightly coloured. They are very secretive forest species, found near small streams. Both are omnivorous

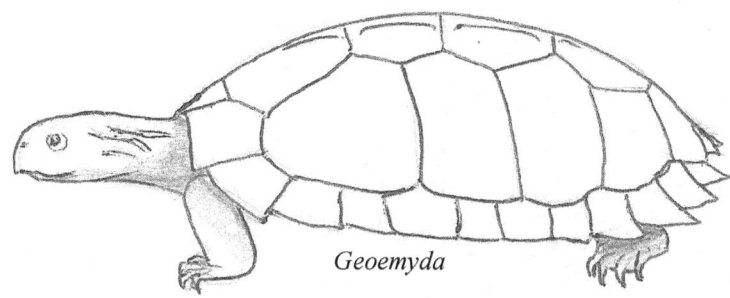

Geoemyda

although the black-breasted leaf turtle can be highly predatory. Supposed *Geoemyda* fossils are known from as far back as 49 million years ago but these are very fragmentary and almost certainly misidentified (e.g. European *G. ptychogastroides* and *G. eureia*, Sri Lankan *G. striata* and Japanese *G. takasago*) and the genus has probably always been restricted to China and Japan. Fossils from the Ryukyu archipelago (the living species and the fossil *G. amamiensis*) are known from around 50,000 years ago. The living species was found on the islands of Okinawa, Kume, Tokashiki and Ie, but is known only as fossils from the latter. Its extinction on Ie and the extinction of the fossil species are probably the result of the fragmentation of the range as the formerly much larger islands shrank with sea-level rise over the past 5 million years. The islands were originally connected to Taiwan and China by a land bridge, but this was lost 1.5 million years ago, from which time the fauna of the islands was isolated, and faced dramatic reductions in area.

The Chinese black-breasted leaf turtle has genetic groups separated at least in part by river courses. There is a distinct genetic group in Vietnam, although that population does not seem to be isolated by any barrier. In China the Guangdong and Guangxi populations are separated by the Xi Jiang River. The Guangxi form is also found on Hainan Island which was joined to that part of China until 2 million years ago.

The *Batagur* group is a south and south-east Asian assemblage of aquatic turtles, ranging in habitats from large rivers and estuaries to forest streams. They include the pairing of *Malayemys* and *Ortilia*, followed by *Geoclemys*, then *Morenia*, the pairing of *Hardella* and *Pangashura*, and the *Batagur* river turtles that characterise the group. These are all highly aquatic animals, with the exception of *Malayemys* which is more specialised than its rather generalist fellows.

Malayemys is restricted to two vulnerable snail-eating turtles: the western

western Malayan species turtle *M. macrocephala* (Burma to Malaysia) and the eastern Mekong species *M. subtrijuga* (isolated parts of Malaysia, Java, Thailand and Vietnam). They are small (20 cm) turtles of wetlands (such as rice paddies) feeding on

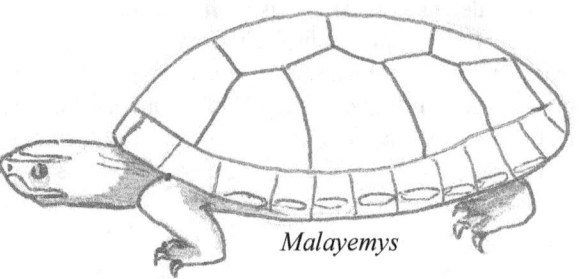

Malayemys

snails. An additional phalanx is present in the 5th digit of the fore-limbs (2-3-3-3-3), the extra-long toes expanding the webbing of this highly aquatic genus. Its relative, *Ortilia* is the Malaysian giant turtle *O. borneensis* of Borneo, Sumatra and Malaysia. It has a scattered distribution in the larger rivers and estuaries. It feeds on a wide range of plant and animal matter, including fish and snakes. Due to its large size (80 cm) it is heavily consumed and considered an endangered species.

Geoclemys, Morenia, Hardella, Pangshura and *Batagur* are a diverse grouping, comprising large river turtles and small species. All members of this highly aquatic group are found from Pakistan to Vietnam. They probably originated around the Indus, Ganges and Brahmaputra rivers of the Indian subcontinent. Fossils of Batagur are known from Nepal 16-12 million years ago, and *Geoclemys* (Indian *G. sivalensis*) and *Pangashura* (Pakistani *P. tatrotia*) from 3 million years ago. They are all highly aquatic and probably

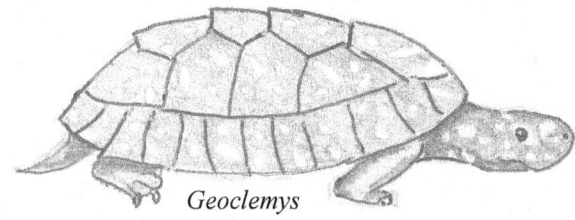

Geoclemys

all share an Indian origin. *Geoclemys* is a monotypic genus, containing only the spotted or black pond turtle *G. hamiltonii* which is found in the Indus and Ganges rivers of Pakistan, northern India and Bangladesh.

It is a large, mainly carnivorous species (35 cm) found in large rives and lakes.

Morenia contains the two small (22 cm) 'eyed turtles' named for their distinctive yellowish eye-like rings on the scutes; the Burmese eyed turtle *M. ocellata* of Burma and possibly China and the Indian eyed turtle *M. petersi* of northern India, Bangladesh and Nepal. Both are considered vulnerable. They are highly aquatic herbivores, found in slow-moving rivers and lakes.

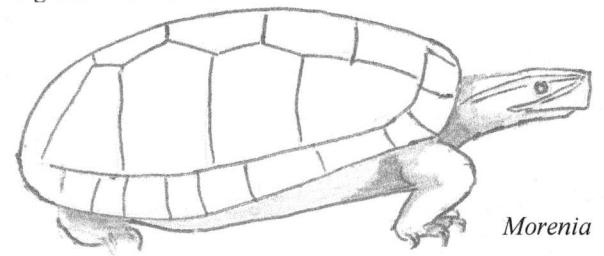

Morenia

Hardella, Pangshura and *Batagur* are large river turtles from south and south-east Asia. The relationship between the genera are not clear; some studies place *Hardella* with *Pangashura*, others group *Batagur* and *Pangashura*.

Hardella contains only the threatened crowned river turtle *H. thurjii* which is found in the Indus river in Pakistan and the Ganges and Brahmaputra in northern India and Bangladesh. The Indus population used to be considered a distinct species,

Hardella

H. indi but this seems to be just a smaller (35 cm instead of 50 cm) geographical form. They are herbivores of slow-moving rivers. A fossil of an aquatic geoemydid apparently similar to *Hardella* has been found in Thailand from 40 million years ago. This is an unexpectedly early Asian geoemydid and may be misidentified.

Pangshura originated around 5 million years ago with fossils of *P. tatrotia* known from Pakistan 3 million years ago. This species may be related to the living Indian roofed turtle *P. tecta* of the Ganges and Brahmaputra rivers, and some other smaller rivers, of north India, Bangladesh, Nepal and Pakistan. The other living species are the endangered Assam roofed turtle *P. sylhetensis*, the brown roofed turtle *P. smithii* and the Indian tent turtle *P. tentoria*. The genus seems always to have been restricted to the large Ganges and Brahmaputra rivers. The brown roofed turtle has a nominate subspecies in most of its range and a northern subspecies, the pale-footed roofed turtle *P. s. pallidipes* in northern India and Nepal. The Indian tent turtle has a nominate subspecies in India and Bangladesh, the pink-ringed tent turtle *P. s. circumdata* in northern India and the yellow-bellied tent turtle *P. s. flaviventer* in north-west India, Bangladesh and Nepal. The brown roofed turtle lives in slow moving waters and is largely vegetarian. The Assam roofed turtle is a carnivorous forest species found in small streams. The Indian tent turtle is found in slow moving large rivers, lakes and is often kept in temple ponds. It is less aquatic than the brown roofed turtle and is largely vegetarian but may feed on meat left by crocodiles. All tent turtles have humped, slightly serrated keeled shells and patterns on the head. The Indian tent turtle has bold markings with yellow stripes on the neck and red patches on the head.

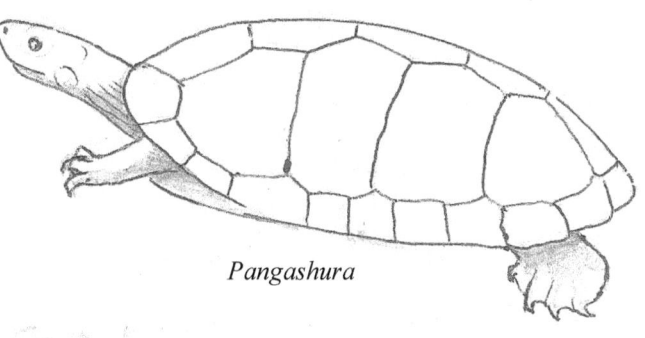

Pangashura

Batagur (including genera sometimes called *Callagur* and *Kachuga*) comprises six large (56 cm) species, all from south and south-east Asia. The most western species are the red-crowned roofed turtle *B. kachuga* from the Ganges and Brahmaputra rivers of north India, Bangladesh, the three-striped roofed turtle *B. dhongoka* from much the same range and the northern river terrapin *B. baska* from isolated estuaries from north India, Bangladesh, Burma, Malaysia, Sumatra and possibly Thailand. The first two species may also occur in Nepal but this is not confirmed. The red-crowned roofed turtle and the three-striped roofed turtle live in the same river system with largely overlapping ranges. The red-crowned is largely herbivorous whereas the three-striped is omnivorous, with a preference for mussels. The range of these species borders the Burmese roofed turtle *B. trivittata* from two isolated parts of the Ayeyarwady and Salween river systems in Burma to the east. The painted terrapin *B. borneoensis* has a more easterly range from Borneo, western Malaysia, Thailand, Sumatra and Borneo. This is a large species (60 cm) found in estuaries and mangroves. Females nest on beaches and the juveniles are able to tolerate sea water for a considerable period of time, this is necessary as it takes time for them to swim to the brackish estuaries. This is one of the most dramatic looking of the Asian turtles; females are generally brown but males have patterned shells and pale heads with a contrasting red stripe on the top of the head. The Malay river terrapin *B. affinis* is found throughout south-east Asia and has distinct western (*B. a. affinis* - Sumatra and Malaysia, possibly as far north as Burma although no longer surviving in Singapore or Thailand) and eastern (*B. a. edwardmolli* - Cambodia and Vietnam to Malaysia and formerly Thailand) subspecies. All species are heavily consumed by humans and threatened with extinction.

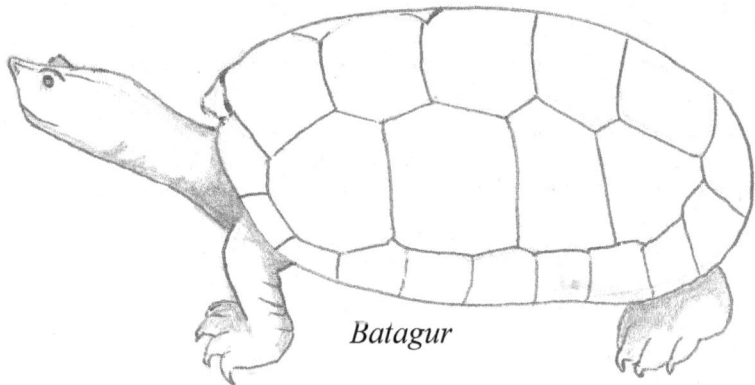

Batagur

The *Cuora/Mauremys* group contains the pairing of *Vijayachelys* and *Melanochelys*, and a poorly resolved grouping that comprises *Sacalia, Leucocephalon, Heosemys, Notochelys, Cyclemys, Mauremys* and *Cuora*. The group ranges throughout Eurasia and includes fully aquatic and semi-aquatic species. The first species are Indian, and the most derived are east Asian. It is probable that the group evolved in the large rivers of the Indian subcontinent, dispersing south-eastwards, and becoming increasingly terrestrial and specialised for forest environments.

144

Vijayachelys and *Melanochelys* are largely Indian genera. They both have carapaces with three keels and heads that are often patterned with yellow and orange markings but are otherwise quite different animals. *Vijayachelys* is a small (13 cm) lowland forest turtle with some degree of plastral mobility. *Melanochelys* is much larger (to 39 cm) and is either a forest species or may be highly aquatic in rivers and marshes

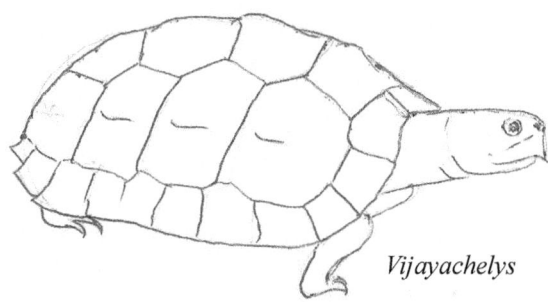

Vijayachelys has only one species, the endangered Cochin forest cane turtle *V. silvatica*. It is an enigmatic species that has been placed with *Geoemyda* and *Heosemys* before finally (in 2006) being given its own genus. This species is restricted to the Western Ghats of India. It was first discovered in 1911 but not recorded again until one was found in 1982. It is now known to be a small (13 cm) easily overlooked and secretive forest species.

Melanochelys includes the tricarinate hill turtle *M. tricarinata* of north-east India, Bangladesh and Nepal, and the Indian black turtle *M. trijuga* of India, Burma, Nepal, Bangladesh and Sri Lanka, with introduced populations in the Chagos and Maldive islands. This species has several subspecies: the nominate south Indian (and possibly Pakistan) subspecies, the Cochin black turtle *M. t. coronata* of Kerala state, India, the Burmese black turtle *M. t. edeniana*, the Bengal black turtle *M. t. indopeninsularis* of Bangladesh, Nepal and eastern India, Parker's black turtle *M. t. parkeri* of Sri Lanka, the Sri Lankan black turtle *M. t. thermalis* of Sri Lanka and Tamil Naud, India. These seem to have diverged around 3 million years ago. The tricarinate hill turtle is a small (16 cm) forest species whereas the Indian black turtle is a large (39 cm) river and lake inhabiting turtle.

Melanochelys

The other members of this group are predominantly east Asian, starting with the genus *Sacalia*. Members of this genus are the endangered 'eyed turtles' of southern China, northern Vietnam and Hainan island: Beal's eyed turtle *S. bealei* and the four-eyed turtle *S. quadriocellata*. These are small turtles (18 cm) with slightly keeled brown shells. They are distinctive in having highly conspicuous 'eye-spots' on the top of the head. In Beal's there are either one or two pairs of eye-spots; in the four-eyed there are always two pairs. Although they both inhabit the same geographical area their precise ranges only overlap on Hainan island. Beal's is a lowland omnivore whereas the four-

Sacalia

eyed is a mountain carnivore species. In addition to these two the Hainan eyed turtle *S. pseudocellata* has been described. This localised form (Hainan island) is thought to be a hybrid between the four-eyed turtle and the golden coin turtle *Cuora trifasciata*.

The divergence of *Sacalia* is probably followed by *Heosemys*, the closely related pair of *Leucocephalon* and *Notochelys*, and *Cyclemys*, although other relationships have been proposed. One alternative is that *Leucocephalon* is not as closely related to *Notochelys* but rather that *Notochelys* groups with *Cyclemys*, and also to *Heosemys*. All arrangements give *Mauremys* and *Cuora* as the most derived members of the grouping. Although *Sacalia* is aquatic the rest of the group are predominantly semi-aquatic.

Heosemys (or *Hieremys*) contains the spiny turtle *H. spinosa* of Malaysia to Indonesia and the Philippines; the endangered yellow-headed temple turtle *H. annandalei* of south-east Asia; and the closely related pairing of the Arakan forest turtle *H. depressa* from a very restricted area of Burma and the giant Asian pond turtle *H. grandis* of south-east Asia. These highly threatened turtles are all forest turtles with flattened, keeled shells. The smallest and most distinctive is the spiny turtle which is 22 cm long and has prominent spines on the marginal scutes and a highly serrated keel on the shell. The Arakan forest turtle is slightly larger (25 cm) and less spiny, but otherwise similar. The giant Asian pond turtle is the most unusual *Heosemys*, with its much greater size. This is one of the largest Asian turtles at 48 cm. They are all forest species found in or near

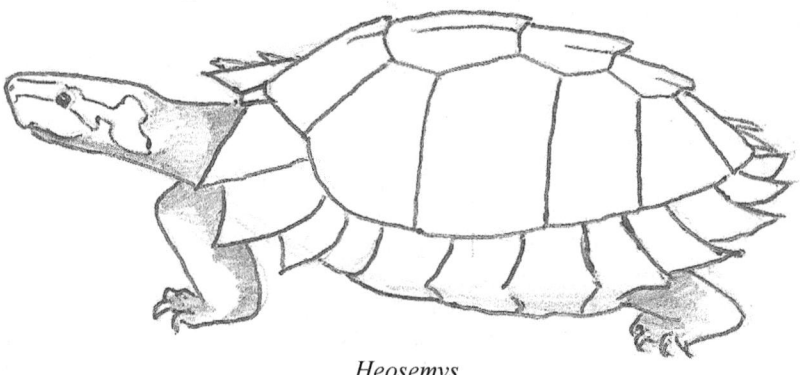

Heosemys

shallow water in forest areas. They are active both in the water and on land and feed in both environments. Most are omnivorous although the spiny turtle is largely vegetarian. Most species seem to represent geographical and size variations of the same basic form, with the only sympatric species being the most different in size: spiny turtle and giant Asian pond turtle. The exception is the yellow-headed temple turtle which does not resemble the others. The yellow-headed temple turtle is found in a small area of Thailand and Vietnam and in an isolated part of Malaysia. This herbivorous turtle is a swamp species, living in damp habitats rather than being truly aquatic. The common name comes from many animals being captured and released in lakes in Buddhist temples, although many others are eaten. A fossil resembling a three-keeled geoemydid resembling *Melanochelys* or *Heosemys* has been found in Thailand from about 45 million years ago; as with other early supposed geoemydids, its identity is uncertain.

Leucocephalon, *Notochelys*, *Cyclemys*, *Cuora* and *Mauremys* probably evolved some 20 million years ago in east Asia. All except *Notochelys* and *Mauremys* are semi-aquatic. The closely related genera *Leucocephalon* and *Notochelys* are both threatened monotypic genera of the south-east Asian islands. *Leucocephalon* is the Sulawesi forest turtle *L. yuwonoi* of a small part of Sulawesi island (Indonesia). *Notochelys* is the Malayan flat-

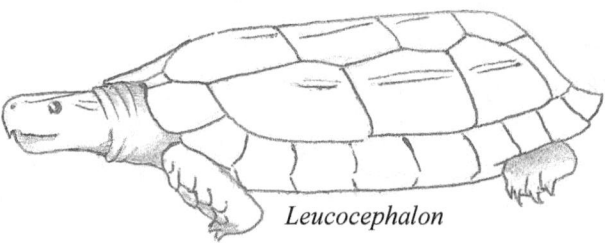

Leucocephalon

shelled turtle *N. platynota* of south-east Asia: Borneo, Java and Sumatra, Malaysia, Singapore and Thailand. This species has a moveable plastron. The Sulawesi forest turtle is found in thick forests and is semi-aquatic whereas the Malayan flat-shelled turtle is found in slow moving water, in lakes and flooding forests. It is more aquatic that the Sulawesi forest turtle but is also very active on land and may be able to make long terrestrial journeys between water bodies. Both are largely herbivorous. *Leucocephalon* probably evolved from a *Notochelys*-like ancestor that

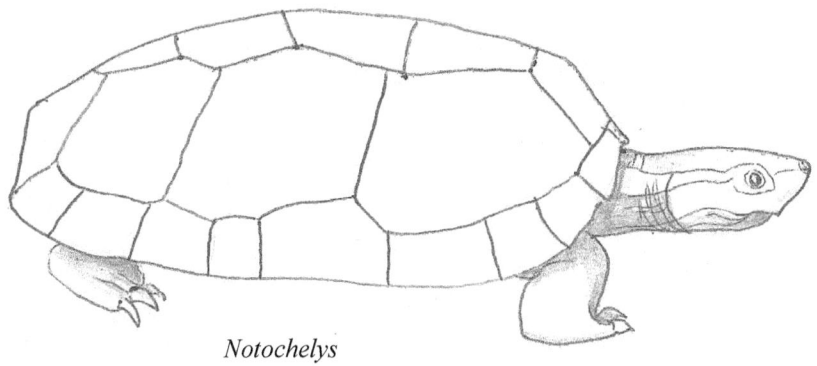

Notochelys

147

crossed the narrow, but deep, marine channel between Sulawesi and the other islands, *Notochelys* was able to spread from mainland Asia into Borneo, Java and Sumatra when low sea-levels united these areas on the Sunda Shelf until 8,000 years ago.

Cyclemys, *Cuora* and *Mauremys* probably originated in east Asia 20 million years ago. The majority are semi-aquatic, with full aquatic specialisation in *Mauremys*. *Cyclemys* are the leaf turtles, semi-aquatic animals found in or near slow-moving waters in forested hills. They are opportunistic in habits and diet and well defended by foul-smelling cloacal secretions; some species also have a hinged plastron. As a result of these defences and their adaptability they are relatively widespread and abundant. The genus is divided into western and eastern groups. The western group are found from Bangladesh to Malaysia and Sumatra: the Assam leaf turtle *C. gemeli* (north India and Bangladesh), Burmese brown leaf turtle *C. fusca* (Burma) and the enigmatic leaf turtle *C. enigmatica* (Malaysia, Sumatra, Java and Borneo). The eastern group is found throughout Indochina from Burma to Vietnam, with Oldham's leaf turtle *C. oldhami* as the original species (a highly variable form ranging from Burma to Thailand), followed by three yellow-bellied species. These are the Vietnamese eastern black-bridged leaf turtle *C. pulchristriata* and the pair of the western black-bridged leaf turtle *C. atripons* (Cambodia) and Asian leaf turtle *C. dentata* (Palawan, Borneo and Java). These species appear able to hybridise as there is a population of possible hybrids between the Burmese brown leaf turtle and Oldham's in Burma.

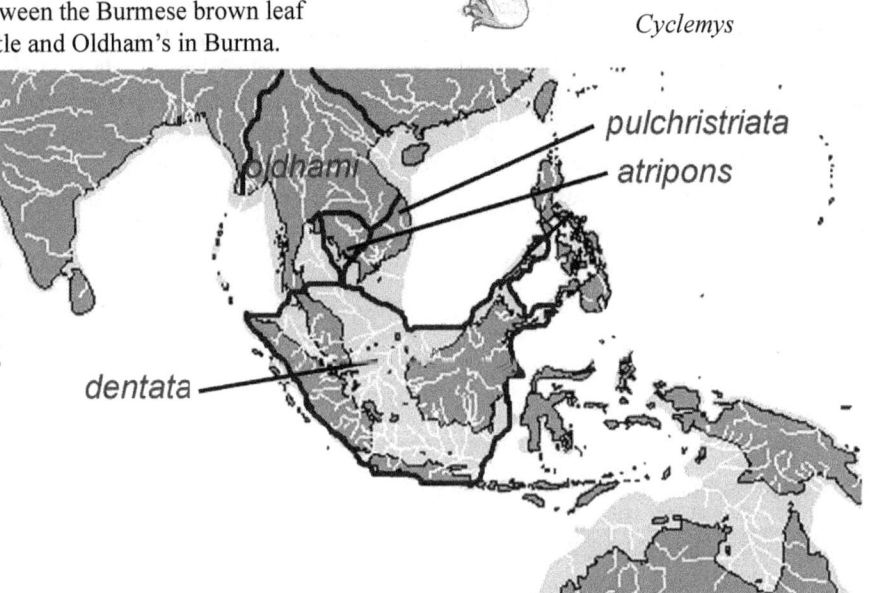

Cyclemys

Eastern species of *Cyclemys* on 'Sundaland' and east Asia 18,000 years ago.

148

It is probable that these *Cyclemys* species were isolated in different parts of south-east Asia over the Ice Ages of the past 2 million years when cold, dry conditions isolated forest patches with the different species being restricted to the core of their ranges. Notably, the enigmatic was probably isolated in Malaysia and the Asian leaf turtle in Palawan. Changing conditions resulted in a lowering of sea levels from 20,000 to 18,000 years ago, uniting the islands on the Sunda Shelf which extends from Malaysia to Borneo and the southern Philippine islands, including Palawan. At this time the shelf would have been dry land and the present day islands would have risen as mountains. This would have enabled the Asian leaf turtle to invade the range of the enigmatic leaf turtle, resulting in hybridisation. 14,600-14,300 years ago rapid sea level rise of more than 5 cm a year flooded the area into the North Sunda River valley between Malaysia and Thailand-Cambodia. From 13,500 years ago the Sunda plain was flooded (now 70m below the surface) isolating Palawan from the other islands. Borneo was isolated from about 10,210 years ago, Java from about 9,300 and Sumatra from about 8,300 years ago. Today the enigmatic leaf turtle is distinctive morphologically and in nuclear DNA but has Asian mitochondria. This indicates that it is a valid species but that all living individuals are descended from a hybridisation between enigmatic fathers and Asian mothers (which provided all the mitochondria present today).

The two most diverse genera are also the most closely related genetically: *Cuora* and *Mauremys*. At first glance these two genera do not seem to be obviously related; *Cuora* are east Asian semi-terrestrial box turtles whereas *Mauremys* are Eurasian river terrapins. *Cuora* are secretive species from forest habitats whereas *Mauremys* are adaptable river animals. The ancestral *Cuora/Mauremys* taxon probably evolved in Asia, separating into the two forms through specialisation for semi-terrestrial and aquatic lifestyles. The oldest representatives of these genera are about 23 million years old.

Mauremys has the widest natural geographical range of any non-marine turtle, ranging from Portugal to Japan. It contains at least nine species and mitochondrial and nuclear DNA indicate that several undescribed species exist. The early European geomeydids have been suggested to be ancestral to *Mauremys*: the earliest genus *Palaeoemys* was followed by *Bergouniouxchelys, Cucullemys, Cuvierichelys, Euroemys, Francellia, Juvemys, Owenemys* and *Provencemys*. The fossils most closely related to *Mauremys* are the French genera *Palaeochelys, Promalacoclemmys* and, in particular,

Mauremys

Palaeomauremys. These are mostly poorly defined and their evolutionary relationships unidentifiable. The spread of this group out of Europe could not have taken place until 52 million years ago because of the presence of the Siberian Sea (or Obik or Ural Sea) between the Ural mountains of Russia and the Polish Lowland Basin (Dniepr-Donetz depression). There may have been a colonisation route at the southern end where this waterway reached the Tethys Sea in an area of shallow water. Fluctuating water levels may have allowed a temporary corridor for colonisation in Kazakhstan at 57 million years ago. By 52 million years ago the Polish Lowland Basin had become isolated from the sea and was freshwater; aquatic terrapins would have been able to make the crossing, but a land connection was not established until 30 million years ago. Although *Mauremys* is identifiable in the European fossil record from 25 million years ago they do not appear in Asia until more recently; fossils from China are no more than 20 million years old at the most, and are followed by the fossil species *M. yabei* and a supposed '*Ocadia*' species (probably a misidentified *Mauremys*) from Japan dating from about 2 million years ago.

The first *Mauremys* fossils from France (*M. massiliensis* and *M. subpyrenaica*) may have been contemporaneous with fossils from Egypt, but are poorly dated, covering a possible range from 23 to 5 million years ago. *M. massilensis* appears to be the ancestral European *Mauremys* and 15 million years ago it gave rise to the central European (Austrian and German) *M. batalleri, M. pygolopha, M. samantica* (including '*Ocadia*' *sophiae*) and *M. steinheimensis*, and the more southern European species. The southern group comprise *M. portisis* of Italy (3 million years ago), which is related to the pair of *M. campanii* (Italy 10 million years ago) and *M. gaudryi* of France (3 million years ago), which are in turn related to the living *Mauremys* species. *M. campanii* fossils are known from an island environment which has an interesting diverse sub-tropical fauna. At 8 million years ago there was a great change in this fauna with a decline in the mammals, including the loss of European primates, but turtles such as *Mauremys* and soft-shell turtles remained.

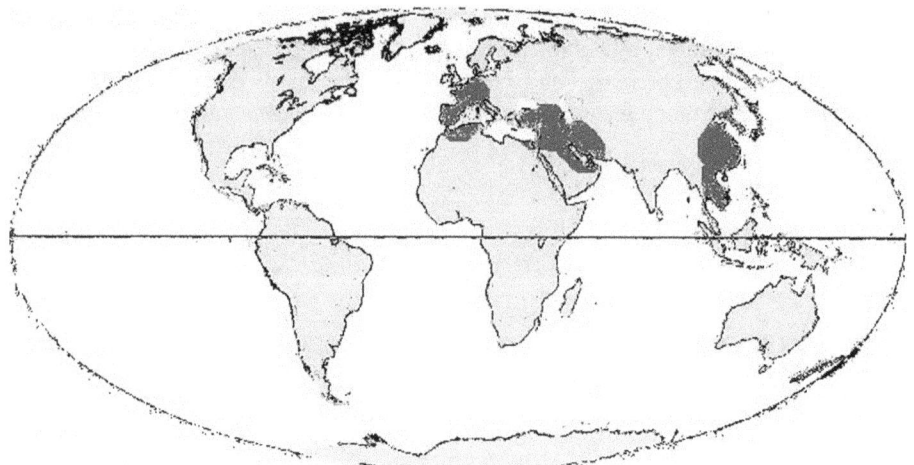

Distribution of *Mauremys* species, living and fossil.

Mitochondrial DNA studies reveal that three genera are contained in *Mauremys*: the typical *Mauremys* species, those described as *Chinemys* and *Ocadia*. These form four groups, two east Asian (*M. japonica* + *Chinemys* + *Ocadia* and *M. annamensis* + *M. mutica*) and two restricted to western Eurasia (*M. caspica* + *M. rivulata* and *M. leprosa*). Of these living species it seems probable that the Mediterranean pond turtle *Mauremys leprosa* may represent the survivor of an ancient lineage which differentiated before the main division of the complex. This lineage dates from 25 million years ago with the species *M. subpyrenaica*, and subsequently *M. rotundiformis* (20 million years ago) and *M. pygolopha* (5 million years ago) and ultimately *M. leprosa* in north Africa, Iberia and southern France. This seems to have given range to isolated populations at the eastern edge of the Mediterranean around 15 million years ago: the Balkan terrapin *M. rivulata* from Bulgaria to Jordan and the Caspian terrapin *M. caspica* which is found from central Turkey to Iran and south to the northern edges of the Arabian peninsula. They spread into east Asia by 20 million years ago when fossils similar to the living *M. reevesi* (in the *Chinemys* group) are known from China.

The Mediterranean pond turtle *Mauremys leprosa* is the most western of all *Mauremys* species, being found in Spain, Portugal and north Africa. It has been present

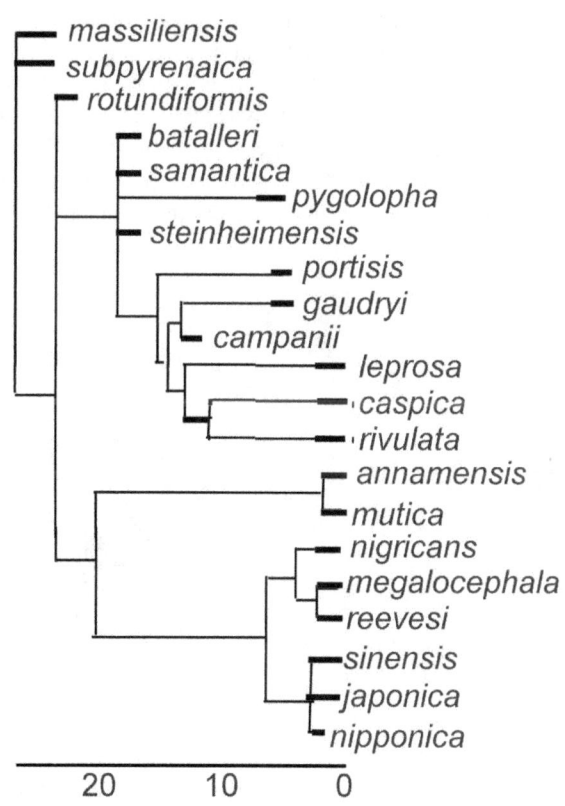

Mauremys phylogeny. Horizontal scale in millions of years ago.

in the area for at least 3 million years, as indicated by fossils from Algeria. Eight subspecies have been described but only two can be supported genetically: *M. l.leprosa* (Iberian Peninsula and northern Morocco) and *M. l. saharica* (southern Morocco, eastern Algeria and Tunisia). These two subspecies have clearly different ranges, in Morocco being isolated by the Atlas Mountains, although one genetic *M. l. saharica* individual has been found north of the Atlas Mountains in Morocco, possibly as a result of dispersal. The greatest genetic diversity is in Morocco and southern Portugal, suggesting that Ice Age refugia were present in these areas. *M. l. leprosa* seems to have spread widely, including crossing the Strait of Gibraltar.

The other European species, the Caspian and Balkan terrapins, have overlapping ranges in the Near and Middle East and South-east Europe. Hybridization is very rare although they are similar and diverged only 6 million years ago. The ancestral species was probably present in western Asia (fossils from 12 million years ago of the Transcaucasus and United Arab Emirates), and expanded into Europe and Turkey. The Balkan terrapin is genetically uniform over much of its range and seems to have spread rapidly from western Asia where it retains most diversity. This species is found on Greek islands and is a successful colonist as it can tolerate brackish water. The Caspian terrapin has more diversity and geographical variation; in the south (Persian Gulf) there is most diversity, with less diversity in the north. This is probably a result of Ice Age restriction to refugia in Central Anatolia, the south coast of the Caspian Sea and the Gulf of Persia, and more recent expansion. There is high diversity in Anatolia despite much of the area being over 1000 m and an unlikely refugium. This diversity does not match the described subspecies (*M. c. caspica* – Mediterranean coast of Arabia to Georgia, *M. c. siebenrocki* - Gulf of Arabia to Turkmenistan, *M. c. ventrimaculata* – Iran). Hybrids have been found in Turkey, where in the breaks in mountain chains allow populations to mix.

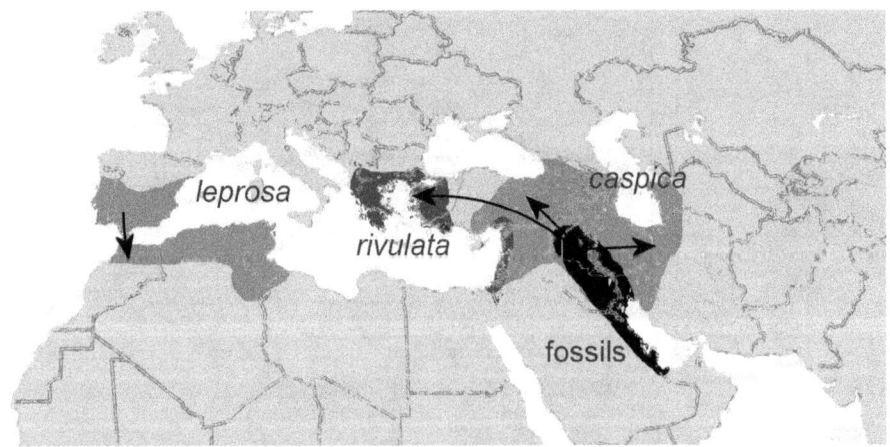

Dispersal of the European *Mauremys* - Mediterranean pond turtle (*M. leprosa*) in the west, Balkan (*M. rivulata*) and Caspian (*M. caspica*) in the east. The range of the fossil ancestors of the eastern species is shown in black.

Asian groupings have a similar long history, with Indian fossils from 10 million years ago. The living Asian group comprises the group of *Chinemys-Ocadia-Mauremys japonica*, and the related pair of the Annam leaf turtle *M. annamensis* and the yellow pond turtle *M. mutica*. This latter pair is probably similar to the early Asian members of the genus. The Annam leaf turtle is restricted to central Vietnam (historical Annam). The yellow pond turtle is found in central and southern China to northern Vietnam and Laos, with populations on Taiwan and Hainan, and a distinct subspecies on the Japanese island of Yaeyama (Ryukyu yellow pond turtle *kami*). The Ryukyu island population would have been able to colonise the islands when they were connected to China via a China-Taiwan-Ryukyus land bridge. This existed from around 2.5 million years ago, being submerged finally 1.5 million years ago. As with all *Mauremys* species they both have camouflaged brown carapaces and a pattern of yellow lines on the head. These are indistinct in the yellow pond turtle but the Annam leaf turtle is the most strongly marked member of the genus and the broad yellowish lines on its head make it resemble a rather dull Southeast Asian box turtle *Cuora amboinensis* from the same region. They are animals of well vegetated swamps and are largely herbivorous.

Within the *Chinemys-Ocadia-Mauremys japonica* group *Chinemys* is distinct from the close pair of the endangered Chinese stripe-necked turtle *M. sinensis* (formerly *Ocadia sinensis*) of coastal southern China to northern Vietnam, and the island of Taiwan and Hainan, and the Japanese pond turtle *M. japonica* from southern Japan. There is also a fossil species from about 2 million years ago are known from Japan ('*Ocadia' nipponica*). The Japanese pond turtle is currently found on Honshu, Shikoku

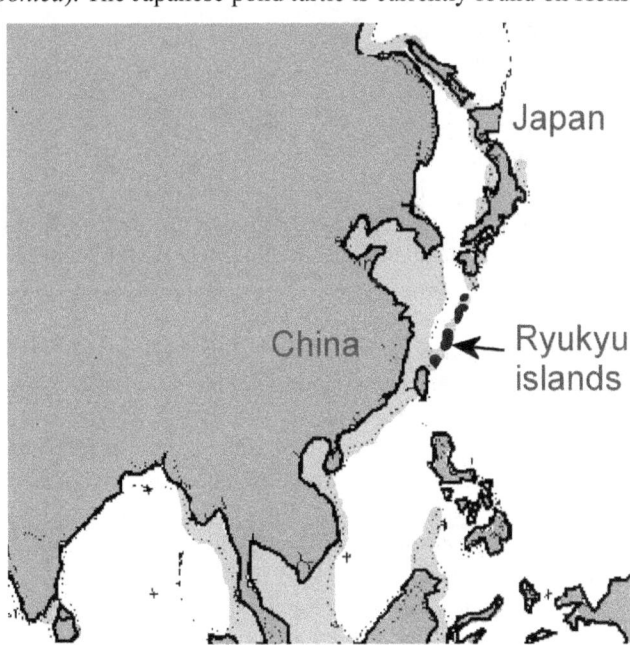

East Asia 2 million years ago showing the Ryukyu islands united with Taiwan and China by dry land, now shallow water.

153

and Kyushu islands. This species declined in range substantially during the last Ice Age, with mitochondrial DNA studies indicating that it was isolated in two small refuges: one in the east in Shikoku island and one in the west in Kyushu island. These tiny remnant populations diverged slightly genetically before conditions improved 10,000 years and they spread back over much of Japan. Both forms colonised Honshu island and have mixed in the centre of that island. The Chinese stripe-necked and the Japanese pond turtle are unusual in the genus in terms of their appearance. The Chinese stripe-necked is one of the most distinctive Asian turtles, with a moderate sized (24 cm) dark carapace contrasting with the skin which is finely striped in yellow, brown and black. The Japanese pond turtle is the most uncharacteristic member of the genus, being reddish brown and lacking the yellow lines on the skin of other *Mauremys* species; it is also unusual in being more of a river species than a marsh one.

'*Chinemys*' contains the red-necked pond turtle *M. nigricans*, and the pair of the Chinese broad headed turtle *M. megalocephala* and Reeve's turtle *M. reevesii*. These endangered species are all basically Chinese species found in isolated streams; the Chinese broad headed turtle in east China; the red-necked pond turtle in southern China and Reeves' turtle China and Korea, with possible prehistoric introductions to Japan and Taiwan and recent introductions to Indonesia, Japan and the Philippines. There are fossils apparently of this species from 20 million years ago.

Cuora are the Asian box turtles. As with the unrelated North American box turtles the plastral lobes are hinged, allowing the turtles to close the shell completely. The earliest fossils of this genus date to 9-5 million years ago from Yunnan province in China (*C. pitheca*). Eleven living species

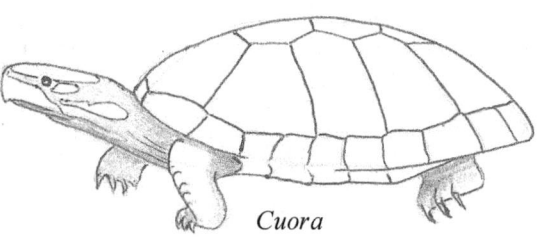

Cuora

have been described in the genus. The earliest to diverge are the south-east Asian box turtle *C. amboinensis* from Bangladesh to the Philippines and the keeled box turtle *C. mouhotii* (formerly *Pyxidea mouhotii*) of China to India and Thailand. The south-east Asian box turtle is widespread but has a fragmented range, probably due to loss of undisturbed marsh habitats. It is found in the Philippines, Sulawesi and the Moluccas (subspecies *C. a. amboinensis*), Sumatra and Java (*C. a. couro*), Thailand, Cambodia, Vietnam and Malaysia (*C. a. kamaroma*) and Burma (*C. a. lineata*). This range is a relict of a more widespread distribution 18 to 20,000 years ago. At that time low sea levels meant that the southern islands of the Philippines, Sumatra, Java and east Asia would have been united as one large land-mass (Sundaland) with only a narrow sea channel to Sulawesi and the Moluccas. Although this channel would have been very narrow, powerful ocean currents have served to prevent most animals from crossing it. This forms 'Wallace's line', the line that marks the north-western limit of the range of Australian animals; to the west of the line Asian animals are found, to the east lie a mixture of Asian and Australian animals. The south-east Asian box turtle is one of the few animals to cross this line; as a semi-aquatic animal it would be able to swim,
154

although the strong currents would still have tended to wash most individuals into the open ocean. It is most probable that animals entering the water (washed into the sea by storms) on the south Philippine coast would be washed south-westwards by the current and deposited on Sulawesi and the Moluccas. The keeled box turtle is similarly scattered with five main isolated populations divided into two subspecies: the nominate form over most of the range and a distinct subspecies in central Vietnam C. m. *oobsti*). Although it is opportunistic in feeding habits it is a largely terrestrial species of forested hill slopes. This results in a scattered distribution, being absent from lowlands and high forests.

The more derived *Cuora* species are all largely Chinese. They form two groups – the *flavomarginata* and *aurocapitata* groups. The *flavomarginata* group (sometimes considered a separate genus: *Cistoclemmys*) comprises two largely Chinese species: the yellow-margined box turtle *C. flavomarginata* and McCord's box turtle *C. mccordi*; and two closely related species from southern China and Vietnam: the Indochinese box turtle *C. galbinifrons* and the Vietnam box turtle *C. picturata*. The yellow-margined box turtle is found in three scattered localities in China (*C. f. sinensis*), on Taiwan (*C. f. flavomarginata*) and on the Ryukyu islands in Japan (*C. f. evelynae*). Fossils of this species are known from Japan (and an unidentified *Cuora* species) from around 100,000 years ago in the Ryukyu archipelago, giving it a wider range in the past than is the case today. The extinct Japanese *Cuora* species probably disappeared due to long isolation of the region and then the islands of the southern part of the range shrinking under rising sea levels in the last few tens of thousands of years.

McCord's box turtle is a semiaquatic species restricted to shallow water in a very small area of Guangxi province in China. The Indochinese box turtle is more widespread, from southern China to northern Vietnam (*C. g. galbinifrons*), central

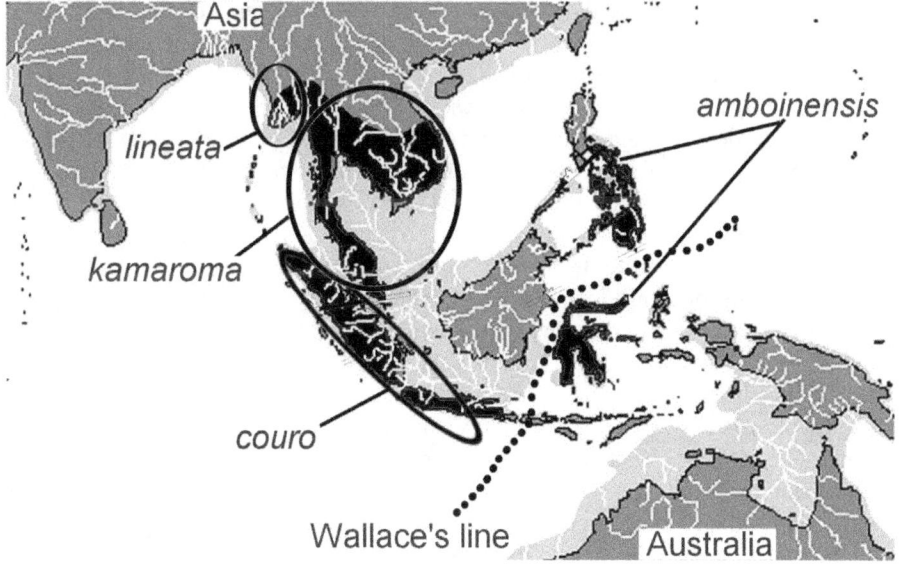

Distribution of *Cuora amboinensis* subspecies, with *amboinensis* the only subspecies to cross Wallace's line.

Vietnam, Laos and Cambodia (*C. g. bourreti*), Hainan island (*C. g. hainanensis*) and southern Vietnam (*C. g. picturata*). Of these *C. g. bourreti* is sometimes recognised as a distinct species. This subspecies differs from the others in reducing the feet by losing a bone in the 4th toe of both fore and hind feet. The toe reduction occurs also in the hind foot of the keeled box turtle and in all limbs of the yellow-margined box turtle which also loses the 5th digit of the hind foot entirely. The reduction in toes is a feature of several semi-terrestrial geoemydids (the above species and *Heosemys spinosa* and *Leucocephalon yuwonoi*), tortoises and the emydid box turtles (*Terrapene*).

The Vietnam box turtle was originally described as a subspecies of the Indochinese but is now recognised as a distinct species. It was known from animals found in markets in Vietnam and was not found in the wild until 2010. The beautifully patterned carapace is perfectly camouflaged in leaf litter and it is remarkably hard to spot in a forest floor. Both the Indochinese and Vietnam turtles are largely terrestrial forest animals. They are primarily predatory, unlike the more omnivorous Chinese species. The yellow-margined box turtle would have been found across China, Taiwan and the Ryukyu islands 2 million years ago when a land bridge connected all three areas. Fossils attributed to the yellow-margined box turtle are known from about 8 million years ago from China (Yunnan), with more recent (1.5 million years ago) fossils from Kyushu and Honshu in Japan, suggesting that it was widespread throughout this region. There are also fossils from Kumejima and Okinawa islands in the Ryukyus which are related, but different from *aurocapitata*. This species was abundant and widespread in the archipelago 1.5 million years ago but disappeared suddenly 100,000 years ago. The land bridge between China, Taiwan and the Ryukyus would have been submerged by rising sea levels 1.5 million years ago, isolating the three living populations and leading to their divergence into distinct subspecies.

The *aurocapitata* group is largely restricted to central-southern China: the Yunnan box turtle *C. yunnanensis* (Yunnan province), Zhou's box turtle *C. zhoui* (no definite locality), Pan's box turtle *C. pani* (no definite locality) and the pair of the golden coin turtle *C. trifasciata* (including *C. cyclornata* as a synonym) and the yellow-headed box turtle *C. aurocapitata*. The Yunnan box turtle was known from 12 museum specimens collected between 1906 and 1908 and was thought to be extinct until 2004 when three individuals were found. Zhou's box turtle is probably from a small area of central China but has not been found in the wild so far. It was described in 1990 based on animals found in Chinese markets. It has been speculated that this species may already be extinct in the wild. It was first bred in captivity in 2004 and it is hoped that a stable captive population can be maintained. Pan's box turtle is restricted to unknown localities in Shaanxi providence. The golden coin turtle is relatively widespread, in coastal south-eastern China and northern Vietnam, with populations on Hong Kong and Hainan. The yellow-headed box turtle is restricted to a very small area of fast-flowing streams and marshes in the valleys of Anhui Province in China; it is not known if this species survives in the wild but it is being bred in captivity. Pan's is largely aquatic in slow-flowing water bodies and is omnivorous. The golden coin is semi-aquatic and largely carnivorous. They are farmed in large numbers, especially on Hainan, although wild populations are endangered. The yellow-headed is omnivorous, with a large vegetarian component.

Recent increases in animal trade in and around China has brought to light a number of rather odd geoemydid (batagurid) turtles, with 13 new species described between 1986 and 2000. Some of these have turned out to be hybrids between familiar species, or even between different genera. The Chinese stripe-necked turtle *Mauremys sinensis* has hybridized with *Cuora trifasciata* to give *Ocadia philippeni* and with *Mauremys annamensis* to give *O. glyphistoma*. *Mauremys reevesii* and *M. mutica* hybridize to give *M. pritchardi*. *M. mutica* hybridizes with *Cuora trifasciata* to give *Mauremys iversoni*. How much hybridisation occurs in the wild is not known but it definitely confuses an already confused situation.

The relationships between these highly threatened *Cuora* species are still muddled and other patterns have been proposed. The scenario described here makes the most sense geographically, but as the paucity of reliable locality data shows, much remains to be discovered and some reinterpretation is inevitable.

References

Artner, H., H. Becker & U. Jost. 1998. Erstbericht über Haltung und Nachzucht der Japanischen Sumpfschildkröte *Mauremysjaponica* (Temminck and Schlegel, 1835). *Emys* **5**: 5–22.

Barth, D. D. Bernhard, G. Fritzsch & U. Fritz. 2004. The freshwater turtle genus *Mauremys* - a textbook example of an east–west disjunction or a taxonomic misconcept? *Zool. Script.* **33**: 213–221.

Chesi, F., M. Delfino & L. Rook. 2009. Late Miocene *Mauremys* (Testudines, Geoemydidae) from Tuscany (Italy): Evidence of Terrapin Persistence after a Mammal Turnover. *Journ. Paleontol.* **83**: 379-388

Claude, J. & H. Tong. 2004. Early Eocene testudinoid turtles from Saint-Papoul, France, with comments on the early evolution of modern Testudinoidea. *Oryctos* **5**: 3-45

Claude, J., E. Paradis, H. Tong & J.-C. Auffray. 2003. A geometric morphometric assessment of the effects of environment and cladogenesis on the evolution of the turtle shell. *Biol. J. Linn. Soc.* 79: 485–501

Diesmos, A.C. J.F. Parham, B.L. Stuart & R.M. Brown. 2005. The phylogenetic position of the recently rediscovered Philippine forest turtle (Bataguridae: *Heosemys leytensis*). *PCAS* **56**(3): 31–41

Ernst, C.H. & J.E. Lovich. 1990. A new species of *Cuora* (Reptilia: Testudines: Emydidae) from the Ryukyu Islands. *Proc. Biol Soc. Wash.* **103**: 26-34

Fèlix, J., J. Budó, X. Capalleras & R. Mascort. 2006. The fossil register of the genera *Testudo, Emys* and *Mauremys* of the Quaternary in Catalonia. *Chelonii* **4**

Feldman, C.R. & J.F. Parham. 2002. Molecular Phylogenetics of Emydine Turtles: Taxonomic Revision and the Evolution of Shell Kinesis. *Mol. Phyl. Evol.* 22: 388–398

Fong, J.J., J.F. Parham, H. Shi, B.L. Stuart & R.L. Carter. 2007. A genetic survey of heavily exploited, endangered turtles: caveats on the conservation value of trade animals. *Anim. Conserv.* (**2007**): 1-9

Frtiz,U., D. Ayaz, J. Buschbom, H.G. Kami, L.F. Mazanaeva, A.A. Aloufi, M. Auer, L. Rifai, T. Silic & A.K. Hundsdorfer. 2008. Go east: phylogeographies of *Mauremys caspica* and *M. rivulata* – discordance of morphology, mitochondrial and nuclear genomic markers and rare hybridization. *J. Evol. Biol.* **21**: 527-540.

Fritz, U., M. Barata, S.D. Busack, G. Fritzsch & R. Castilho. 2006. Impact of mountain chains, sea straits and peripheral populations on genetic and taxonomic structure of a freshwater turtle, *Mauremys leprosa* (Reptilia, Testudines, Geoemydidae). *Zool. Script.* **35**: 97–108.

Fritz, U., G. Fritzsch, E. Lehr, J.-M. Ducotterd & A. Muller. 2005. The Atlas Mountains, not the Strait of Gibraltar, as a biogeographic barrier for *Mauremys leprosa* (Reptilia: Testudines). *Salamandra* **41**: 97-106.

Fritz, U., D. Guicking, A. Auer, R.S. Sommer, M. Wink, M. & A.K. Hundsdörfer. 2008. Diversity of the Southeast Asian leaf turtle genus *Cyclemys*: how many leaves on its tree of life? *Zool. Script.* **37**: 367–390.

Fritz, U., A. Petzold & M. Auer. 2006. Osteology in the *Cuora galbinifrons* complex suggests conspecifity of *C. bourreti* and *C. galbinifrons*, with notes on shell osteology and phalangeal formulae within the Geoemydidae. *Amphibia-Reptilia* **27**: 195-205

Fujita, M.K., T.N. Engstrom, D.E. Starkey & H.B. Shaffer. 2004. Turtle phylogeny: insights from a novel nuclear intron. *Mol. Phyl. Evol.* **31**: 1031-1040

Gong, S., H. Shi, Y. Mo, M. Auer, M. Vargas-Ramírez, A.K. Hundsdörfer & U. Fritz. 2009. Phylogeography of the endangered black-breasted leaf turtle (*Geoemyda spengleri*) and conservation implications for other chelonians. *Amphibia-Reptilia* **30**: 57-62

He J., T. Zhou, D.-Q. Rao & Y.-P. Zhang. 2007 Molecular identification and phylogenetic position of *Cuora yunnanensis*. *Chinese Science Bulletin.***52**(17):2085-2088

Hervet, S. 2003. *Le groupe Palaeochelys sensu lato–Mauremys dans le contexte systématique des Testudinoidea aquatiques du Tertiaire d'Europe occidentale. Apports à la biostratigraphie et à la paléobiogéographie.* Unpublished Ph.D. thesis, Muséum national d'Histoire naturelle, Paris.

Hirayama, R., N. Kaneko & H. Okazaki. 2007. *Ocadia nipponica*, a new species of aquatic turtle (Testudines: Testudinoidea: Geoemydidae) from the Middle Pleistocene of Chiba Prefecture, central Japan. *Paleontol. Res.* **11**: 1-19

Joyce, W.G. & T.R. Lyson. 2010. *Pangshura tatrotia*, a new species of pond turtle (Testudinoidea) from the Pliocene Siwaliks of Pakistan. *Journ. Syst. Palaeontol.* **8**: 449-458

Le, M. & W.P. McCord. 2008. Phylogenetic relationships and biogeographcial history of the genus *Rhinoclemmys* Fitzinger, 1835 and the monophyly of the turtle family Geoemydidae. *Zool. J. Linn. Soc.* **153**: 751-767

Le, M., W.P. McCord & J.B. Iverson. 2007. On the paraphyly of the genus *Kachuga* (Testudines: Geoemydidae). *Mol. Phyl. Evol.* **45**: 398-404

Mantziou, G., N. Poulakakis, P. Lymberakis, E. Valakos & M. Mylonas. 2004. The inter- and intraspecific status of Aegean *Mauremys rivulata* (Chelonia, Bataguridae) as inferred by mitochondrial DNA sequences. *Herpetol. Journ.* **14**: 35-45.

Parham, J.F., W.B. Simison, K.H. Kozak, C.R. Feldman & H. Shi. 2001. New Chinese turtles: endangered or invalid? A reassessment of two species using mitochondrial DNA, allozyme electrophoresis and known-locality specimens. *Anim. Conserv.* **4**: 357–367

Parham, J.F., B.L. Stuart, R. Bour & U. Fritz. 2004. Evolutionary distinctiveness of the extinct Yunnan box turtle revealed by DNA from an old museum specimen. *Proc. R. Soc. B: Biol. Lett.* **271**(1556[S6]): 391-394

Praschag, P., C. Schmidt, G. Fritzsch, A. Müller, R. Gemel & U. Fritz. 2006. *Geoemyda silvatica*, an enigmatic turtle of the Geoemydidae (Reptilia: Testudines), represents a distinct genus. *Organisms, Div. Evol.* **6**: 151-162

Ren, H., K. Naotomo & O. Hiroko. 2006. Fossil Turtles from the Kiyokawa Formation of the Shimosa Group (Pleistocene) at Chiba Prefecture. *Quatern. Res.* **45**: 179-187

Sasaki, T., Y. Tasukawa, K. Takahashi, S. Miura, A.M. Shedlock & N. Okada. 2006. Extensive morphological convergence and rapid radiation in the evolutionary history of the family Geoemydidae (old world pond turtles) revelaed by SINE insertion analysis. *Syst Biol.* **55**: 912-27

Shi, H., J.J. Fong, J.F. Parham, J. Pang, J. Wang, M. Hong & Y.P. Zhang. 2008. Mitochondrial variation of the "eyed" turtles (*Sacalia*) based on known-locality and trade specimens. *Mol. Phylogenet. Evol.* **49**(3): 1025-1029.

Spinks, P.Q., R.C. Thompson & H.B. Schaffer, 2009. A reassessment of *Cuora cyclornata* Blanck, McCord and Le, 2006 (Testudines, Geoemydidae) and a plea for taxonomic stability. *Zootaxa* **2018**: 58–68

Spinks, P.Q., H.B. Shaffer, J.B. Iverson & W.P. McCord. 2004. Phylogenetic hypotheses for the turtle family Geoemydidae. *Mol. Phyl. Evol.* **32**:164-82

Stuart, B.L. & J.F. Parham, 2004. Molecular phylogeny of the critically endangered Indochinese box turtle (*Cuora galbinifrons*). *Mol. Phyl. Evol.* **31**: 164–177

Suzuki, D. & T. Hikida. 2011. Mitochondrial phylogeography of the Japanese pond turtle, *Mauremys japonica* (Testudines, Geoemydidae). *J. Zool. Syst. Evol. Res.* **49**: 141-147

Takahashi, A., H. Otsuka & R. Hirayama. 2004. Fossil Asian box turtle, *Cuora* sp. from the latest Pleistocene of Okinawa Island, Japan. In: *Program of the 153rd Regular Meeting of the Palaeontological Society of Japan*

Takahashi, A., H. Otsuka & H. Ota. 2008. Systematic Review of Late Pleistocene turtles (Reptilia: Chelonii) from the Ryukyu Archipelago, Japan, with special reference to paleogeographical implications. *Pacific Science* **62**(3): 395-402.

Wink, M., D. Guicking & U. Fritz. 2001. Molecular evidence for hybrid origin of *Mauremys iversoni* Pritchard et McCord, 1991, and *Mauremys pritchardi* McCord, 1997 (Reptilia: Testudines: Bataguridae). *Zool. Abh. Staatl. Mus. Tierk. Dresden* **51**: 41-49

13. Tortoises

Tortoises (Testudinidae) first appear in the fossil record of Asia around 60 million years ago. The earliest forms are difficult to identify beyond being fairly large, nondescript, semi-terrestrial or terrestrial turtles; they may be geoemydids. Some of the earliest probable tortoises belong to the genus *Achilemys*. *A. allabiata* is known only from one fragmentary specimen from 45 million years ago in Wyoming, but the 37 cm long *A. cassouleti* from France 53 million years ago is much more complete. These were primitive for Testudinidae, sharing many characters with geoemydids. *Hadrianus majusculus* from North America (New Mexico) 50 million years ago is more clearly a tortoise, rather than a geoemydid. Molecular clock calculations suggest an origin of the family about 70 million years ago, which fits with the earliest fossils dating from about 56 million years ago. With the exception of the very earliest stage, the family Testudinidae has an exceptionally good fossil record, due to the solid skeleton of its members and the fact that many species live in burrows, which makes them prone to being buried in fossilizing conditions. Many of the early genera however are poorly preserved and poorly dated shells of uncertain affinities: Asian *Kansuchelys* (from somewhere between 55 and 25 million years ago), *Sinohadrianus* (45 million years ago except for *S. sichuanensis* which is of uncertain status and may not be a testudinid), *Miotestudo ibba* (Sri Lanka around 15 million years ago) and *Sharemys* (30 million years ago). *Dithysternon valdense* from Switzerland 35 million years ago is perhaps more interesting than most of these fossils, but its most notable feature, a double hinged plastron, is most similar to the hinges seen in some emydid terrapins, rather than the weak single hinge of some tortoises.

The family probably originated in Asia around 70 million years ago and remained restricted to that continent until around 60 million years ago. Suggested testudinoids from North America from before this time, such as *Gyremys spectabilis*, do not seem to be true testudinoids and this example is more likely to belong to the extinct families Bothremydidae or Baenidae. Similarly *Clemmys backmanni* is a macrobaenid. Around 60 million years ago cool conditions enabled many animals (mammals as well as tortoises) to migrate over land-bridges between the northern continents. The early North America fossil tortoises were mostly in the genus *Hadrianus*. The exact identity of this poorly defined genus remains unclear but it shows the trans-Beringean (east Asia to North America) range expected by the first tortoise to move out of Asia. This genus was known from North America from around 50 to 45 million years ago and from Europe around 35 million years ago (Spain). Four possibly valid species have been recorded: *H. majusculus, H. robustus, H. tumidus* and *H. utahensis. Cymotholcus* was formerly placed in *Hadrianus*; it contains *C. schucherti* from Alabama 35 million years ago, which differs little from true *Hadrianus*.

Fossil data suggest that most living tortoise genera originated by 24 million years ago. At this time the global climate was relatively warm and tropical zones widespread. This may have favoured diversification of the Testudinidae in North America, Asia, Europe and Africa. South America remained isolated and was unlikely

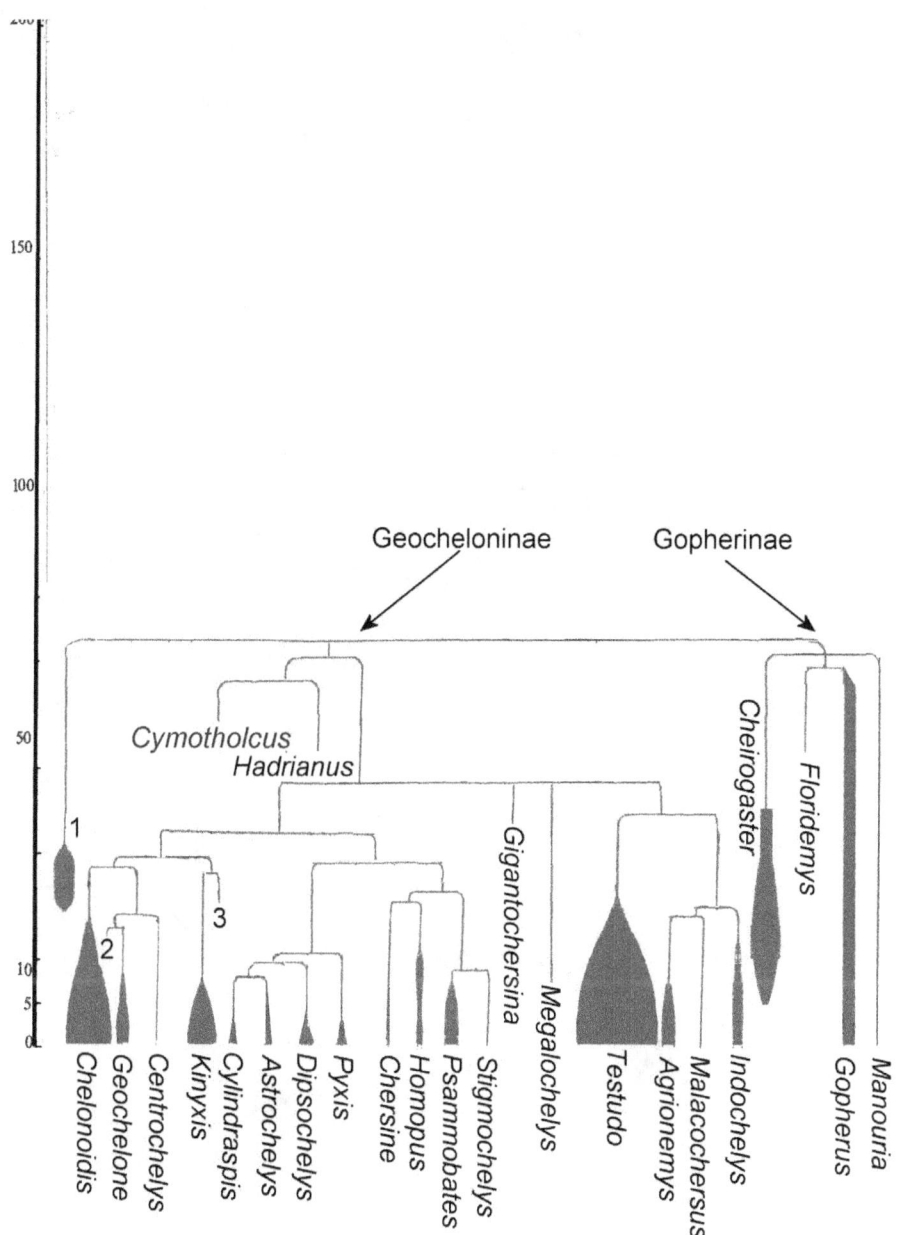

Phylogeny of tortoises
1 - *Ergilemys*; 2 - *Miotestudo*; 3 - *Impregnochelys*

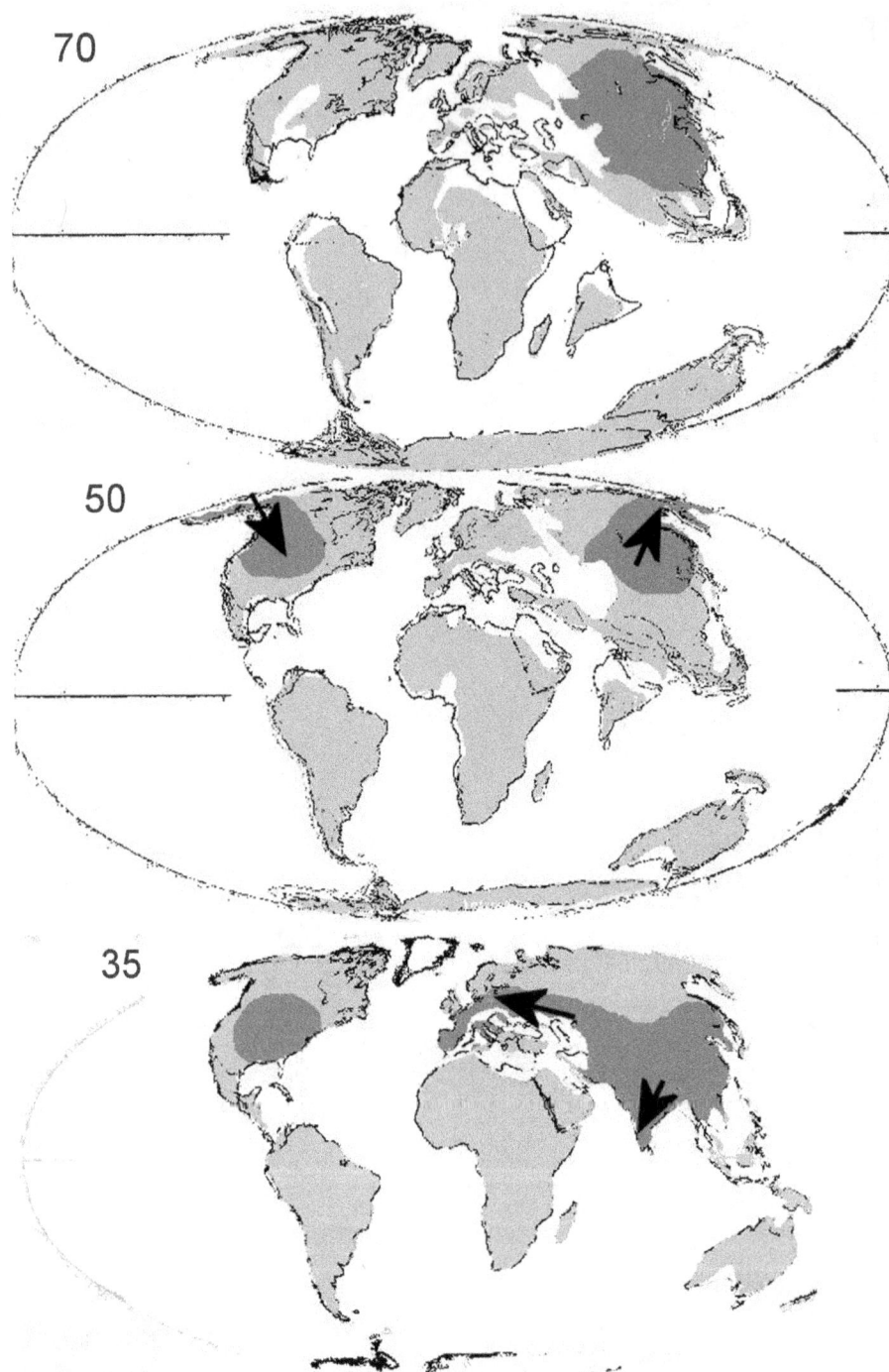

Tortoise distribution from 70 million years ago to 35 million years ago.

to be colonised until 5 million years ago. Australia remains isolated today and has still not been occupied by terrestrial tortoises.

Many early Eurasian '*Hadrianus*' have also been placed in the living Asian genus *Manouria*. The first of these is '*Testudo' eocaenica* from Germany 45 million years ago, which may be ancestral to true *Manouria*. Most of the early tortoises, such as *Ergilemys* from Europe and Asia 25 million years ago, appear similar to *Manouria* and the North American *Gopherus*. At much the same time, fossil tortoises are known from Europe and Asia, many attributed to the genus *Stylemys* (which is similar to *Gopherus*) and to *Manouria*. These are all members of the subfamily Gopherinae, but this is largely in the retention of relatively primitive features, shared with the Geoemydidae. Ancestral characters retained by the Gopherinae and shared with the Geoemydidae and Emydidae include the retention of a non-enlarged pygal bone and the presence of mental glands on the underside of the chin in *Gopherus* (mistakenly reported from *Manouria* as well). These chin glands secrete odours during courtship and encounters between the highly aggressive males.

Other tortoises in Asia at this time seem to be ancestral to the other group of living tortoises, the Geocheloninae. Molecular studies suggest that Gopherinae and Geocheloninae split about 48 million years ago.

Gopherinae

Gopherinae today comprise the genera *Manouria* and *Gopherus*. The two genera are genetically quite distinct and are thought to have been separated for about 45 million years, since the separation of western Asia and North America by rising sea-levels opening the Bering straits.

Many of the Eurasia fossils are of doubtful identity. The Georgian *Manouria vialovi* (53 million years ago, 40 cm) is similar to the German '*Testudo' eocaenica*. There are also fossils of an un-named species from Austria (10 million years ago). *M. obailiensis* 50 million years ago in Kazakhstan is similar to *H. utahensis*. Questionable species include: ?*Testudo margae* (2 million years ago, Sarawak) and *Testudo punjabiensis* (5 million years ago, India). One particularly well preserved *Manouria* species is *M. oyamai* from the Ryukyu archipelago of Japan. This species was present on the islands around 29,000 years ago, having colonised from China and the southern Ryukyu islands across the Kerama Strait which would have been an area of dry land at that time. It disappeared abruptly 23,000 years ago; the cause of this extinction is not known but may be associated with the arrival of humans as the earliest human remains from the Ryukyus are 32,000 years old and by 20,000 humans had spread throughout the archipelago.

Manouria has two living species: the Asian forest tortoise *M. emys* and the impressed tortoise *M. impressa*. These range from Burma to Vietnam and Borneo. The impressed tortoise is the less well known of the two; it is found in mountain forests from Burma eastwards. In these forests its beautifully patterned shell is well camouflaged in the leaf-litter. Its habits in the wild are very little known but it seems to be a specialised species feeding mainly on mushrooms, although it will eat plants such as bamboo shoots. The larger Asian brown tortoise has two subspecies: *M. e. emys* from southern Thailand, Malaysia,

Manouria

Sumatra and Borne and *M. e. phayrei* from north-west Thailand to north-east India. These differ in the rather obscure feature of the pectoral scutes being well separated in *M. e. emys* and joined in *M. e. phayrei*. The Asian brown tortoise is unique amongst turtles in constructing a nest of vegetation. The female scrapes together a large mound of leaves and lays a clutch of up to 50 eggs in the warm centre of this compost heap. She defends the nest site from other tortoises and possibly also from potential predators. In some areas this species tends to feed in water, rather like a geoemydid. This raises the possibility that this is a primitive way of feeding for tortoises, but adaptations of the tongue and the throat musculature show that the Asian forest tortoises have taken to semi-aquatic feeding independently – gopher tortoises are probably better models for the ancestral tortoise.

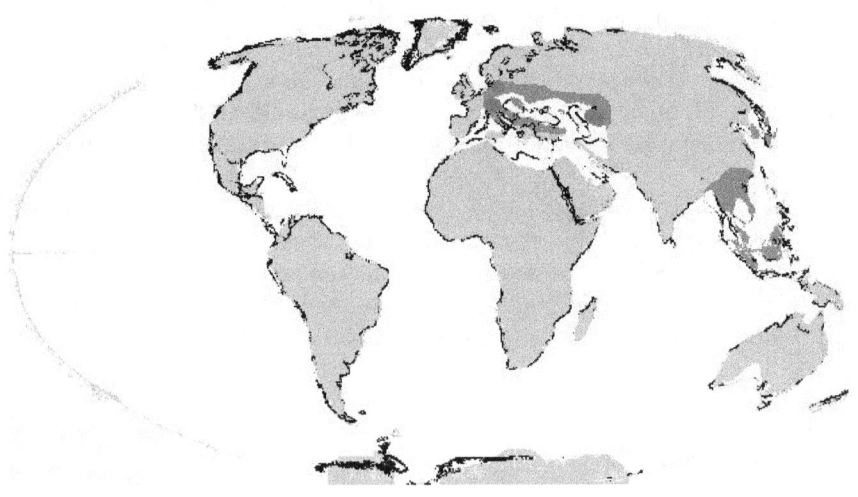

Distribution of *Manouria*; ancient species (before 10 million years ago) in west Eurasia and recent (within the past 2 million years) in east Asia.

164

The gopher tortoises *Gopherus* and *Stylemys* form a 'chronospecies', that is a lineage which can be seen changing through time from one species to another. This is unusual in that it requires a very good fossil record to have an almost complete sequence of evolutionary change. *Stylemys* and *Gopherus* differ in features such as the shape of some of the scutes and the absence of pits in the bones joining the plastron and carapace in *Stylemys*. These are associated with the axillary musk glands found in most tortoises, although the ducts do not run through the shell in other species. However, these characters are often individually variable and the bridge pits develop with age. Thus, there are no clear differences between the early species (*Stylemys*) and the later ones (*Gopherus*). *Hesperotestudo* probably also belongs to this group, as the earliest stage of this chronospecies. At least 16 '*Stylemys*' species have been described but many are probably the same. The oldest remains date from 56 million years ago, but the best fossils are those of *S. nebrascensis* (Wyoming 30 million years ago and Nebraska 34 million years ago) and fossils from the White River Group of the Badlands National Park. These fossils are now mostly identified as *Gopherus laticuneus*, which is the earliest species considered to be a true *Gopherus* rather than a '*Stylemys*'. They date from between 37 and 27 million years ago, when global climates changed, generally from rain forests to cooler, more semi-arid conditions. Living *Gopherus* are associated with arid environments, and the evolution of the genus at this time may be due to specialisation for this environment.

The majority of *Gopherus* fossils are poor; the best preserved have complete skeletons, including well preserved skulls in *G. brevisternus, G. canyonensis, G. laticuneus* and *G. mohavetus*. Analysis of these and the living species seems to indicate that the genus originated with *G. laticuneus* 30 million years ago in central USA (Colorado, Wyoming and Nebraska). This was a giant form (at least 50 cm long) with an elongate skull and well-developed ear region, typical of *Gopherus*.

Gopherus

Around 25 million years ago the *Gopherus* lineage divided into two which correspond to the living '*Xerobates*' and true *Gopherus* groups. Gopher tortoises probably moved from the central states into the southwest, subsequently dividing into east and west forms; by 20 million years ago North America would have had gopher tortoises in the south-west ('*Xerobates*' or *Scaptochelys*) and in the central states (*Gopherus*). Of these *Gopherus* is the more specialised: it has an inflated inner ear region, containing a large otolith (a stone in the inner ear involved in detection of head movements); short, robust neck vertebrae with an interlocking joint at the base of the neck; modified fore-limbs with broad, spatulate digits and limited digit mobility. *Gopherus* uses the limbs in digging and at the same time uses the head and neck to brace against the burrow sides to obtain maximum power. The large otolith has been suggested to be involved in detection of vibrations through the ground; otoliths are normally associated with balance rather than hearing but there is some evidence that some sound detection can occur through this vibration. Even if it otoliths do transmit sound an enlarged one would not give particularly increased sensitivity so what the function of an enlarged otolith is remains obscure. *Xerobates* lacks these specialisations. Today *Xerobates* comprises the western and central North American desert tortoises and true *Gopherus* comprises the central and eastern gopher tortoises.

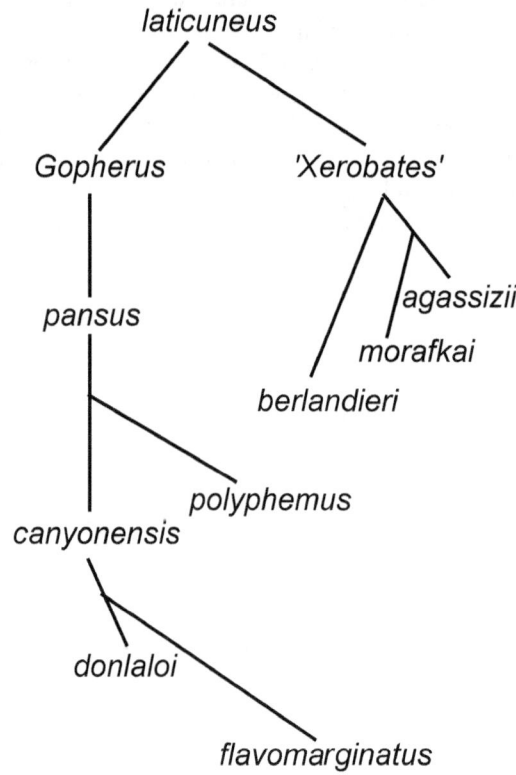

Phylogeny of *Gopherus*.

The *Xerobates* lineage seems to have divided into distinct species 17-19 million years ago, which predates the formation of large deserts in North America by around 12 million years. This seems surprising as the living species of this group are generally considered desert animals. Of the '*Xerobates*' species, Agassiz's desert tortoise *G. agassizii* is known from the past 4 million years, although the deserts it inhabits in the west of the USA (California, Nevada, Utah, Arizona, and fossils from New Mexico) have only existed for the past 7,000 years. Although Agassiz's is a desert-associated species, many of its apparent desert adaptations may be older characters that it retains. These include a slow metabolic rate and high tolerance of extreme fluctuations in water balance; large, hard-shelled eggs; herbivory; and a varied diet with a large digestive tract (maximising efficient digestion of poor quality food). They are burrowing species and may have evolved to burrow in semi-arid grasslands rather than in true deserts. Agassiz's desert tortoise also inhabits woodland and grassland areas (as did fossils of this species). In 2011 it was discovered that the Colorado River formed a barrier to movement of desert tortoises, resulting in genetic divergence between the northern form (*G. agassizii*) and the southern form. This latter form has been named Morafka's desert tortoise *G. morafkai* and is found in scrub and dry forest east and south of Colorado river from Arizona into Mexico. Agassiz's desert tortoise is found in arid scrub to the north-west of the river. Both species feed on grass and, when available, wildflowers and cactus fruit. Agassiz's dig long (up to 7 m) burrows in valleys while Morafka's mainly hide in rock crevices and dig shorter burrows in hillsides. Agassiz's burrows provide a stable, cool and relatively moist environment in which to avoid extremes of summer heat and winter frosts. Burrows may be used for several years and may be used by more than one tortoise; some burrows have been estimated to have been used by tortoises for 5,000 years. Morafka's desert tortoises live in areas with warmer winter temperatures, making deep burrows less important. The other '*Xerobates*' species Berlandier's tortoise *G. berlandieri* is known from fossils from around 1 million years ago and it is found today in a very small area in the south-east of North America (Texas, north Mexico, and fossils from Arizona).

The true *Gopherus* group comprises several fossils as well as two living species. The relationships between the earliest true *Gopherus* are impossible to determine and several species seem to have diverged 10-20 million years ago. These comprise the central USA *G. pansus* (Colorado), *G. vagus* (Wyoming) and *G. brevisternus* (Wyoming). The latter was a short-snouted, wide-skulled species, although less so than the more recent species. It had an expanded ear region and a medium sized ototlith but retained the 'batagurine process' in contact with the pterygoid (a projection on the basioccipital at the base of the skull seen in the Geoemydidae and presumed to be primitively present in tortoises). In almost all its features it is intermediate between *G. laticuneus* and the advanced *Gopherus* species. This then seems to be ancestral to the living *G. polyphemus* (southern Gulf states, 1.5 million years ago to the present) and a group of species comprising the living *G. flavomarginatus* (Mexico and fossils from Texas, from 1.5 million years ago to the present) and the fossils *G. canyonesis* (Texas and Arizona, 3 million years ago) and *G. donlaloi* (Mexico to Texas, 100,000 years ago). Although *G. brevisternus* is the most informative of the early true *Gopherus* it may well

167

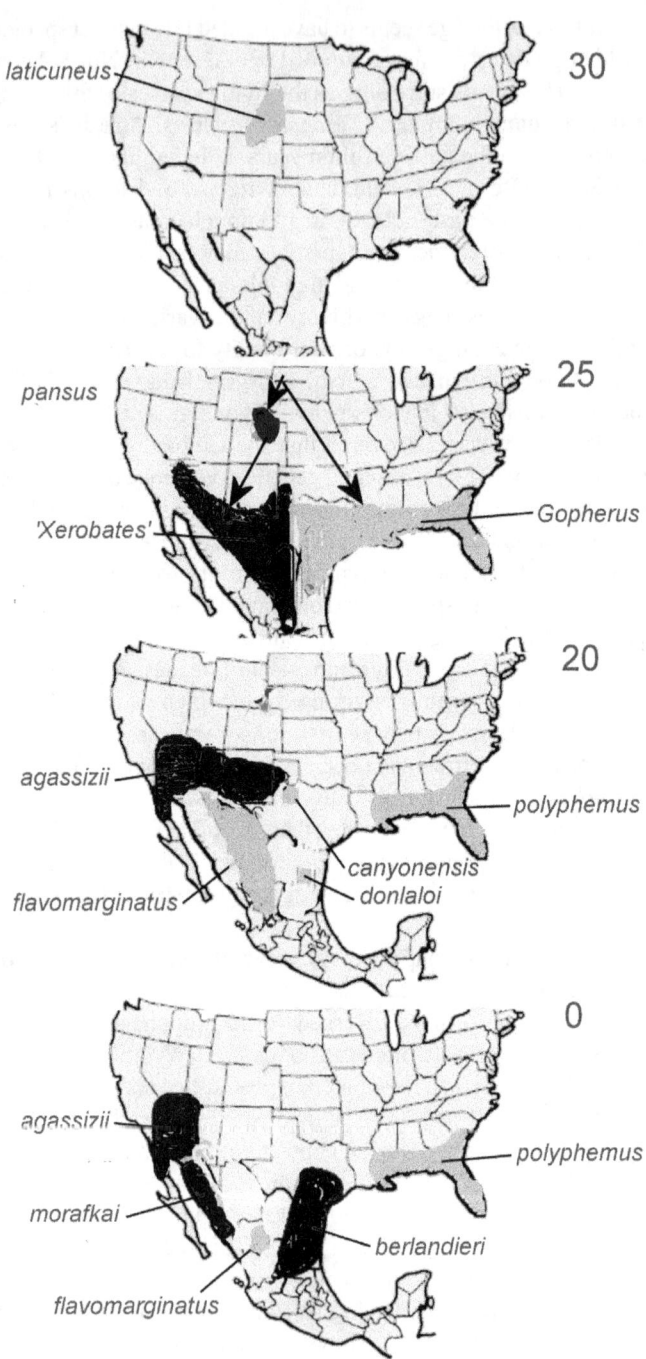

Distribution of *Gopherus* from 30 million years ago, '*Xerobates*' species shown in black.
168

turn out that *G. pansus, G. vagus* and *G. brevisternus* are all the same thing; if that is the case then the name for this species would be *G. pansus* (which was the first name to be applied even though the fossil is not very useful). This group would have originated from the ancestral *G. laticuneus*, moving south into a wide area from Mexico to Florida as conditions became cooler. The centre became too cool for tortoises 10 million years ago, by which time gopher tortoises were limited to the south. Of the living true *Gopherus* the Mexican giant gopher tortoise *G. flavomarginatus* may have been isolated from the other species by the Trans Pecos Texas and Sierra Madre Oriental mountain chains. *G. flavomarginatus* groups with the last two giants, *G. canyonensis* and *G. donlaloi*. It seems that these three species are also really the same lineage, which decreased in size over time, with the 80 cm *G. canyonensis* 3 million years ago in Texas and Arizona, diverging into an inland form, *G. donlaloi* (58 cm) and a more coastal form, the living *G. flavomarginatus* (40 cm). The extinction of *G. donlaloi* may have allowed *G. flavomarginatus* to move into its range, sandwiched between the ranges of the *Xerobates* species. The eastern part of the *Gopherus* range is occupied by *G. polyphemus*. Although this is called the 'desert tortoise' it actually inhabits a variety of habitats.

Gopher tortoises now exist in a small part of the range inhabited by this type of tortoise millions of years ago. Ecological changes have restricted them to the present day arid zones and human impacts are further affecting them. Native American tribes in the south-west USA exploited them for food in the past; this has now been replaced by other mortality factors. A respiratory disease is having a major impact on populations west of the Colorado River, in Nevada and Utah. Habitat has been lost to housing and road developments, agriculture and mining. In addition there has been significant mortality caused by off-road driving in the desert, predation by ravens whose populations have been inflated by the food available in rubbish dumps, and there has been heavy collecting from the wild for the pet trade. As a result populations have declined dramatically and all five species are threatened.

Ancient gopher and gopher-like tortoises were not all giants; there is a fossil dwarf tortoise that has been suggested to be related to this group. *Floridemys nanus* from Florida is known from a single shell from sometime between 11 and 5 million years ago (so contemporaneous with some of the giant gophers) and a poor specimen from 32 million years ago. This latter specimen is so much earlier that it is almost certainly a misidentification. *Floridemys* appears to have been only 10 cm long, far smaller than any of the living species. Exactly where this fits in is unknown.

The North American fossil genus *Hesperotestudo* (or *Caudochelys*) probably also belonged to the Gopherinae. This included several species from 30 to 5 million years ago, with one species surviving to 100,000 years ago in Central America. At least one Chinese species has been referred to this genus but this is probably a misidentification. A typical example is *Caudochelys tedwhitei* from Florida 21-16 million years ago. These had some protective bones in the skin of its fore-limbs which sets it aside from the living North American tortoises which all belong to the genus *Gopherus*. They also differ in having narrower nuchal and pectoral scutes, more femoral-inguinal contact, no premaxillary ridge and a transverse ridge on the jaw. Despite these differences

Hesperotestudo seem to be more closely related to *Gopherus* than to the living South American tortoises (*Chelonoidis*); *Hesperotestudo* and *Gopherus* share extensive ridges on the upper jaw, an elongate vomer bone in the palate and a similar arrangement of the shoulder.

Hesperotestudo was widespread in North America from 16 million years ago. Some 18 species have been described; while some are represented by largely complete skeletons others are known from very poor material, such as *H. kalganensis* which is known from a singe dermal scale. Early species (around 15 million years ago) are recorded from central USA: Nebraksa (*H. angusticeps* and *H. orthopygia*), Kansas (*H. orthopygia*) and South Dakota (*H. niobraensis*). By 5 million years ago they ranged from Kansas and Oklahoma (*H. riggsi*) to Florida (*H. alleni*) and Texas (*H. johnstonia*). Until around 5 million years ago the genus was restricted to North America but then spread southwards and disappeared from its original range. As cold Ice Age conditions eliminated the more northerly populations, *Hesperotestudo* was restricted to the southern states by 2.5 million years ago (unidentified species from New Mexico, *H. turgida* from Texas, *H. incisa* from Georgia). While the gopher tortoises are largely arid habitat species (although now found in dry forests as well as grassland and semi-deserts), *Hesperotestudo* may have been a more forest associated tortoise. As dry conditions expanded in the southern part of North America, favouring the gophers, *Hesperotestudo* may have retreated southwards. Around 100,000 years ago they were restricted to *H. crassiscutata* ranging around the Gulf of Mexico, from Florida to El Salvador, and *H. wilsoni* in Texas, New Mexico and Mexico. This was the last mainland species to survive, lasting to around 50,000 years ago. One exception to this pattern is the late species, *H. percrassa*, from Pennsylvania around 1 million years ago. This very fragmentary species may represent a temporary recolonsation of the northern range during a warm interglacial period. At around this time island populations appeared, such as *H. bermudae* which was present on Bermuda 310,000 years ago. This was a moderate sized tortoise, about 30 cm long. It probably floated to Bermuda from Florida on the Gulf Stream as Florida supported two lineages of *Hesperotestudo* at this time.

Hesperotestudo seems to have fallen victim to climate change, with rising sea-levels eliminating the island populations 90,000 years ago and the North American species being forced southwards, to extinction in Florida and into Central America. The cause of extinction in Central America is not known; whatever the cause, some 50,000 years ago the genus was extinct. When Bermuda re-emerged from the sea 3,000 years ago no potential sea-crossing tortoises remained in the area.

One more genus may be related to the Gopherinae, although its exact position is far from clear. The European fossil genus *Cheirogaster* comprised at least 11 species from 35 million years ago to 3 million years ago, with a peak of diversity around 20 million years ago. These giant fossils (up to 2 meters in *C. perpiniana*) have been suggested to be related to the Mediterranean tortoises (*Testudo*), or to the African spurred tortoise *Centrochelys sulcata*, but it retained a number of primitive features and is most likely to have been a relict of the early tortoises otherwise only known from around 50 million years ago. It seems that 50 million years ago the early tortoise originated in North America and Asia, and the Gopherinae spread across North America

170

Distribution of *Hesperotestudo* from 15 million years ago, showing the movement southwards and the colonisation of Bermuda 100,000 years ago.

(*Stylemys-Gopherus-Hesperotestudo*) and Asia (*Manouria*). The closure of the Obik Sea seaway between Europe and Asia around 35 million years ago enabled the group to spread into Europe, giving rise to the western European *Cheirogaster*. It is notable that geomydid terrapins crossed in the opposite direction some 5-10 million years earlier, when shallow water and temporary land bridges enabled semi-aquatic animals to cross. Tortoises seem to have had to wait until a permanent land route had formed. By 33 million years ago *Cheirogaster* was present in Austria and France, spreading to Spain by 15 million years ago.

In addition ot the Gopherinae (the giant *Cheirogaster*) Europe was occupied by Geocheloninae in the form of the small *Testudo* and related genera. The giants were reasonably successful with populations throughout south and central Europe, north Africa and on several Mediterranean islands. They declined from at least 10 million years ago, last being recorded in France 18 million years ago. They survived in the southern extremity of Europe (Spain) until 4 million years ago. Island populations survived until 2.5 million years ago (Minorca and Malta). Their final disappearance coincided with the start of the last Ice Age. Conditions in Europe would have changed from at least 3 degrees warmer than at present, to cold and dry - inhospitable for giant tortoises.

Geocheloninae

All other tortoises are placed in the subfamily Geocheloninae. This group probably originated around 50 million years ago in Asia when so-called '*Testudo*' fossils are recognised. The early fossils are hard to identify with any certainty and these should be regarded as early Geocheloninae rather than being consigned to any specific genus.

Geocheloninae can be divided into two groups: the 'testudonans' (*Testudo* and its relatives, *Malacochersus* and *Indotestudo*) and all remaining Geocheloninae. It seems that the earliest Geocheloninae evolved in Europe and west Asia and dispersed into Africa following the closing of the Tethys seaway between African and Eurasia, which presented a barrier to colonisation of Africa until about 40 million years ago. The oldest African fossil tortoise is *Gigantochersina ammon* from 35 million years ago which seems to have been the first African member of the Geocheloninae. The earliest fossils of this genus are from Egypt where several different forms have been described as different species, now all regarded as belonging to a single species. Something similar has also been found in Kenya; these fossils are attributed to '*Gigantochersina ammon*' but are only 4 million years old, and so may not be the same. These poorly known giants were not restricted to Africa as at around the same time *Megalochelys atlas* and *M. cautleyi* lived in India (with Javan remains also attributed to the former); this giant probably attained a length of nearly 2 m.

After the appearance of *Gigantochersina* there is a gap in the African fossil record until 20 million years ago; whether tortoises remained rare in that dark period or we simply lack the fossils is not known. By 20 million years ago they had diversified and we find abundant African fossil tortoises. Most of these are doubtfully identified as '*Geochelone*' and '*Testudo*'. In reality these genera have been used for fossils largely in the sense of a large tortoise (*Geochelone*) or a small tortoise (*Testudo*) so tend to be misleading at best. They seem to have divided into two clear groups: a group of largely small southern African endemics and the larger tortoises (probably including *Gigantochersina*). The former comprises lineages leading to the living *Chersina, Homopus, Stigmochelys* and *Psammobates*. Whether of not all southern African fossil tortoises belong with this group is far from clear.

Fossils suggest that *Homopus*, *Chersina* and *Stigmochelys* evolved between 20 and 12 million years ago, with the living species originating within the last 5 million years ago. It is probable that the ancestors of these small to medium-sized tortoises were found throughout the grasslands of southern Africa until 14 million years ago. At that time the cold Benguela current off the south-west coast of Africa developed, causing a dramatic change in the region's climate. This shifted cloud patterns northwards, leading to a drying of southern Africa from 14 to 11 million years ago. Semi-arid and desert areas spread, isolating the tortoises in the Cape Province where the meeting of cold and warm ocean currents maintained a milder, wetter climate. These isolated populations became increasingly specialised to the prevailing dry conditions, leading to the southern African tortoises being small and mostly flattened (especially in *Homopus*). *Homopus* and *Chersina* are genetically extremely close, although morphologically distinct.

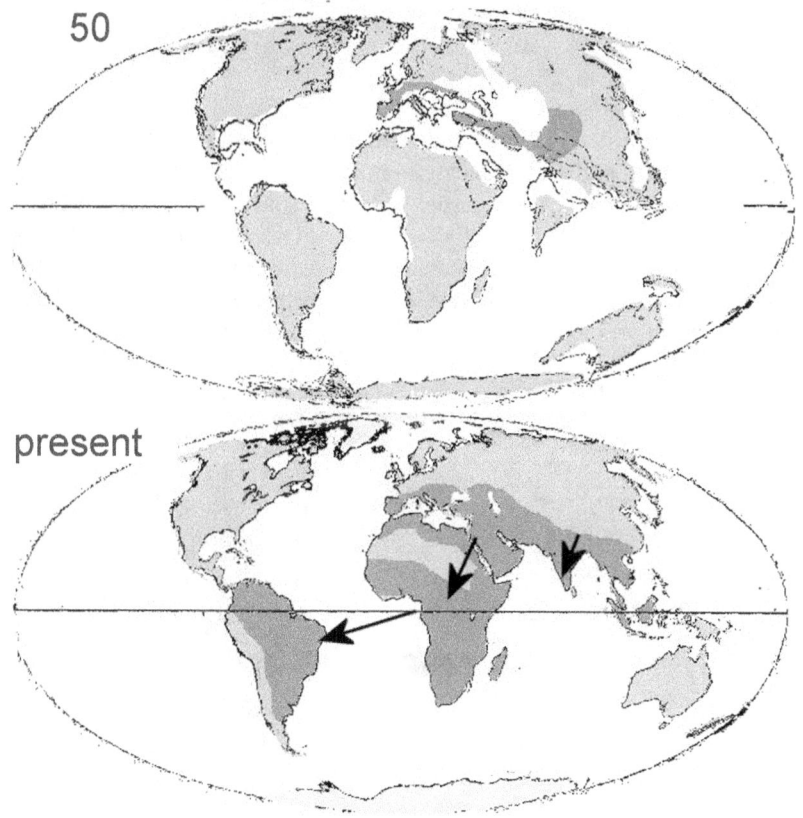

Distribution of Geocheloninae from 50 million years ago.

Homopus contains two species groups; true *Homopus* and a group sometimes called *Chersobius* (speckled padloper *H. signatus,* Nama tortoise *H. solus* [formerly *H. bergeri*] and the Karoo dwarf tortoise *H. boulengeri).* The true *Homopus* are the parrot-beaked tortoise *H. areolatus* and the greater dwarf tortoise *H. femoralis.* The '*Chersobius*' group differ from others *Homopus* in having five claws on the forelimbs instead of four and having males having a concave plastron. In addition they have a single inguinal scute and 11-12 marginals. Two subspecies (*signatus* and *cafer*) have been described in the speckled padloper based largely on shell colour, but these do not seem to be valid. The *signatus* subspecies does form a group but is contained within *cafer*, and the subspecies may reflect local adaptations to different coloured rocks. The Karoo dwarf and the speckled padloper are sister species. The Namaqua tortoise had a confused history: first recognised in 1955 it was initially considered to be a Karoo dwarf but was then attributed to '*H. bergi*' although this species was very obscure and is now thought to be a misidentification of the tent tortoise *Psammobates tentorius*. This confusion was clarified only recently, in 2007 when it was named *H. solus*.

Homopus

There is little evidence from which to interpret the origins of *Homopus* as their association with arid environments and small size means that fossils are rare. In fact the only fossil *Homopus* is a partial body cast (rather than a fossil of the actual animal) from an unknown date in the eastern Cape. Despite its poor quality this fossil has been named, as *H. fenestratus*. The genus seems to have always been restricted to the southern tip of Africa; the living species are all restricted to the Cape except for the Nama tortoise from Namibia and the greater dwarf which extends into the north-west of South Africa.

Chersina

Homopus

woods & savannah

Psammobates

desert

semi-desert

grassland

Distribution of Southern African tortoises.

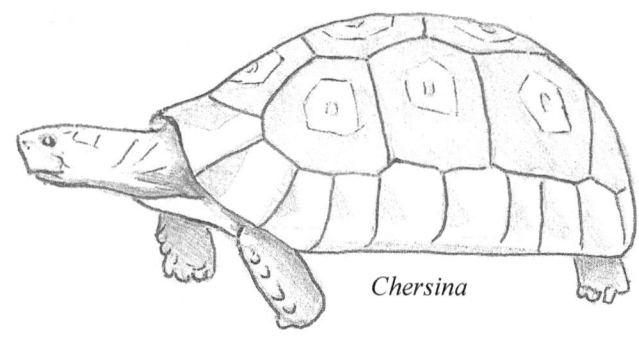

Chersina

The angulate tortoise *Chersina angulata* is the only member of its genus. It diverged from *Homopus* about 18 million years ago according to molecular clock calculations, so the separation of the two genera may have preceded the dramatic drying of the region. It is possible that they diverged as habitat specialists, with the padlopers adapting to more arid conditions and the angulate preferring scrub habitat. The early *Chersina* relative from that split is *Mesochersus orangeus* from Namibia. This occurred from 17 million years ago to 5 million years ago and was followed by true *Chersina* 4.5 million years ago. The angulate tortoise is restricted to the southern tip of Africa and is morphologically relatively uniform across its range but contains two genetic groups (as defined by mitochondrial DNA). These groups are from the western and southern Cape. These areas correspond to refugia that would have survived the drying event. The development of the Benguela current would have led to the development of a pronounced east-west climate gradient, with droughts becoming frequent in the west. The populations isolated in the west of the species' range adapted to increasing summer droughts, whilst the southern area remained relatively stable and damp. The range of both populations probably contracted as coastal areas became increasingly inhospitable and the tortoises survived in small pockets in mountain valleys. In the western population this fragmentation led to further divergence which is reflected today by genetically distinct north-western and south-western populations.

Psammobates is a very restricted genus, comprising the tent tortoises of southern Africa. All three species are similar in appearance, being under 15 cm with yellow shells and radiating black star-like patterns on the scutes, this gives them a superficial similarity to the Indian starred tortoise *Geochelone elegans*, and in describing the first specimen Linnaeus confused the two. The genus appears to have always been restricted to semi-arid habitats, mainly in South Africa. They seem to be somewhat intermediate between padlopers and the angulate tortoise in their habitat adaptations; some are scrub species like the angulate but others have adapted to more arid environments. They divide into the semi-arid scrub associated South African tent tortoise *P. tentorius* which is found from southern Namibia to the western Cape and the two closely related specialists of even more arid habitat. The semi-arid habitat occupied by the South African tent tortoise is dividied into three areas, occupied by three slightly different subspecies: the southern *tentorius* from the Cape, the west-central *trimeni* from north-west South Africa to south-west Namibia and the eastern *verroxii* from central

Namibia to the Great Karoo. Of the arid specialists the serrated tortoise *P. oculiferus* is the most widespread, being found throughout the desert areas of southern Africa. This species lives in arid savannah and woodland patches in these apparently inhospitable deserts. It is very cryptic and little known. The geometric tortoise *P. geometricus* is the most restricted, being found only in a tiny area

of the Western Cape Province of South Africa. It lives in the very specialised renosterveld which is a mixture of bushes and sand scrub, in particular renosterbos ('rhinoceros bush'), wild olives and rosemary. The climate is extreme with warm summers and cool winters. It hibernates in cool seasons and is inactive for a large part of the year. It has a very slow rate of reproduction, nesting only once a year and laying 2-4 eggs. This is partially compensated for by the fast growth and early maturity of the juveniles. Even so, the population of this species is estimated at only 4,000 animals.

The only member of this group not to be small and restricted to southern African arid habitats is the leopard tortoise. *'Geochelone' pardalis* has been placed in several different genera; its inclusion in *Geochelone* is incorrect as it shares only a relatively large size with other members of that genus; it has a close genetic relationship with the southern African tent tortoises and is sometimes placed in *Psammobates* but it is morphologically so different that the most sensible arrangement is to recognise it as its own genus: *Stigmochelys*. This species is the most widespread African tortoise, ranging from the Horn of Africa to South Africa. Two subspecies have been described based on shell shape and colour: *S. p. pardalis* has been recorded from the south-west of the range, and *S. p. babcocki* from elsewhere. This is not supported by DNA studies and this

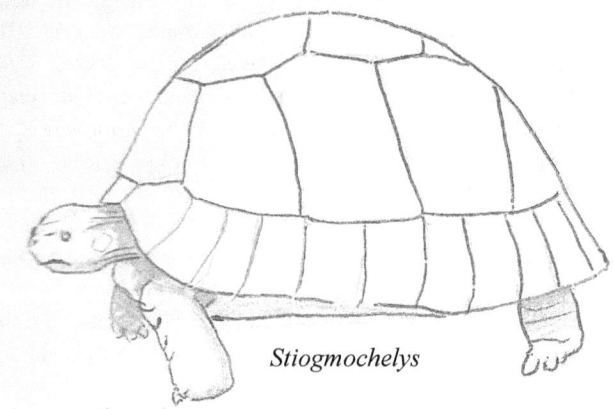

Stiogmochelys

seems to be a single wide ranging species varying in appearance geographically but not forming distinct subspecies. The greatest genetic diversity is found in southern Africa, probably due to restricted gene flow during the Ice Ages when arid climates would have fragmented the range, particularly in the south of Africa.

This genus appears to be closely related to some of the many poorly defined '*Geochelone*' species of east and southern Africa. The earliest species suggested to be in this lineage is '*Geochelone*' *namaquensis* (now referred to *Namibchersus namaquensis*) from Namibia about 15 million years ago, followed by '*Geochelone*' *stromeri* of South Africa and '*Geochelone*' *crassa* from Kenya 4 million years ago. There are giant forms suggested to be *Stigmochelys* species from Ethiopia from around 3 million years ago. The earliest definite *Stigmochelys* fossils are the Kenyan *S. brachygularis* which range from about 4 million years ago to 2.5 million years ago. There are relatively abundant fossils comprising shells, bones and eggs, and these were very similar to the living leopard tortoise in appearance and size. The leopard tortoise appears in the fossil record 3 million years ago in South Africa. East African fossils are slightly more recent, at 2 million years ago (in Kenya and Tanzania). They seem to have evolved in the grassland regions of east and southern Africa and, although their range has changed as habitats have moved, leopard tortoises have remained little changed over 20 million years.

The larger tortoise group is diverse and corresponds to most of the old concept of '*Geochelone*' with the addition of the small *Kinixys* and *Pyxis* tortoises. The ancestor of this group was probably a large tortoise similar to one of the poorly defined fossils labelled '*Geochelone*'. By 31 million years ago it was well established in Africa and had started diverging into the living species. This form had considerable dispersal potential, with its large size enabling it to survive prolonged periods of food shortage and to cross areas of poor habitat or float across water bodies. Floating across water allowed Madagascar to be colonised about 20 million years ago, giving rise to the Western Indian Ocean radiation of at least 12 species (see Chapter 15). The tortoises that dispersed over land appear to have specialised into a small forest or scrub form, the hinge-backed tortoises *Kinixys*, and the larger forms of North Africa. This latter group (true *Geochelone*) dispersed eastwards into India, either over land through Arabia or by drifting across the Arabian Sea. African *Geochelone* was probably the ancestor of the South American tortoises (Chapter 16), again through drifting on ocean currents. Genetic studies are divided, some suggesting that the South American *Chelonoidis* are related to *Geochelone* and others suggesting a relationship to *Kinixys*. The small size of *Kinixys* makes a poor candidate for crossing oceans, whereas *Geochelone* would be able to survive a long sea-crossing with comparative ease.

The hinge-backed tortoise of the genus *Kinixys* have an unusual hinge in the carapace between the 4th and 5th costals and 7th and 8th peripherals. This enables the back of the carapace to be lowered, protecting the hind limbs. This is unusual in two ways: no other group of turtles has a hinge in the carapace; the hinge also protects the hind limbs rather than the more normal protection of the head. This might suggest that the hinge evolved in a burrowing or crevice-living ancestor where the head was protected by the surroundings but the hind-limbs were exposed; however, Kinyxis does not look like a burrower in other respects. They are known as fossils from 19 million years ago in east Africa, but are

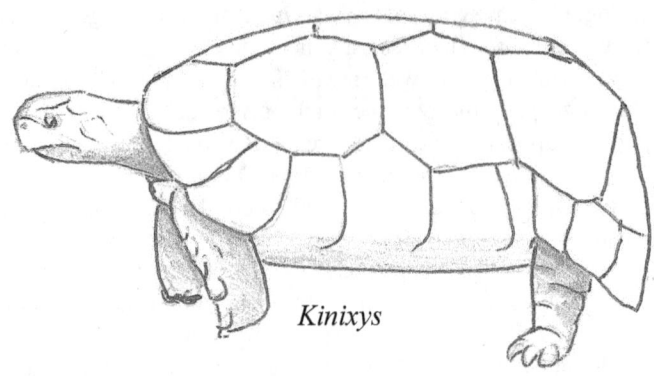

Kinixys

estimated to have originated about 30 million years ago. The Kenyan fossil species *Impregnochelys pachytectis* from 18 million years ago may be a *Kinixys* relative. This is the only mainland African genus to have diverged significantly outside of the small southern African group; it has several species and subspecies, most of which have probably arisen rapidly as adaptations to a changing environment. There was rapid aridification of East Africa following the closing of the Indonesian seaway from around 4 million years ago. This interrupted the flow of tropical Pacific water into the east Indian Ocean, reducing the cloud generated by the meeting of ocean currents and its transport across to east Africa. The history of the genus is probably one of adaptation and movement with changing habitats. For much of its history hinge-backed tortoises seem to have been mainly east African, but by 5 million years ago they had reached South Africa.

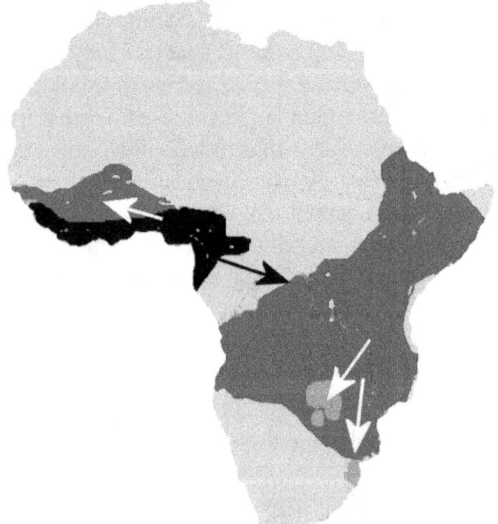

Range expansion of *Kinixys*; black - early range (*homeana* and *erosa*), dark grey - *belliana* and eastern species (*spekii*), light grey - *lobatsiana* and *natalensis*.

Although the earliest fossils are east African from 19 million years ago the ancestral *Kinixys* species may have been a west African species similar to Home's hinge-back *K. homeana* and the serrated hinge-back *K. erosa*. These are the oldest living species and originated in dry woodland and savannah environments of northern Africa. Bell's hinge-back *K. belliana* is more widespread and prefers more arid environments, as do the southern Natal hinged tortoise *K. natalensis*, Speke's hinge-back *K. spekii* amd the Lobatse hinged tortoise *K. lobatsiana*.

Geochelone tortoises appear to have evolved some 30 million years ago. Fossils from 20 million years ago have been identified as being related to the African spurred tortoise *G. sulcata* (sometimes placed in its own genus, *Centrochelys*). These have also been suggested to be fossils of the leopard tortoise genus *Stigmochelys* but as they are from North Africa and Arabia, covering the range of the living spurred tortoise the *Centrochelys* identification seems most probable. The earliest fossils cannot be identified with any confidence, but fossils from 18 million years ago indicate that *Centrochelys* was found throughout northern Africa and Arabia (whether one species as today, or several species, is not known). A similar form '*Geochelone*' *burchardi* from the Canary islands around 5 million years ago must have floated across from west Africa, a crossing of 100 km. Another poorly known giant from the same time, '*Geochelone*' *robusta*

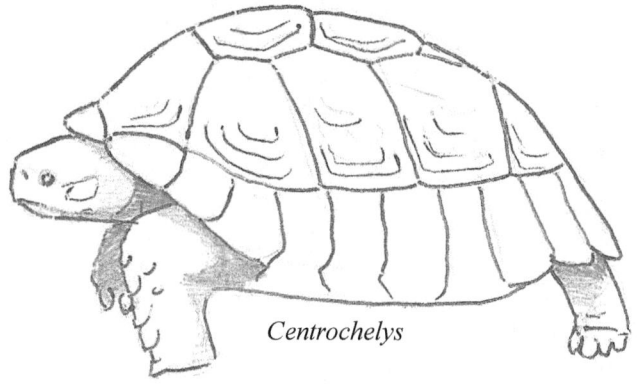

Centrochelys

of Malta, might also have been related to these giants that were spread throughout the dry savannah and semi-desert that spread across north, east and southern Africa. The living African spurred tortoise *C. sulcata* is found in such arid environments, where deep burrows enable it to hide from the hottest times of day. These animals are well adapted to the harsh conditions of this part of Africa, where they feed on coarse grasses of low nutritional quality. In captivity a much richer diet results in extremely rapid growth, highlighting the adaptability of some tortoise species. The range of the African spurred tortoise is fragmented into seven separate populations. There is a major division between the eastern and western populations which form distinct genetic groups. These are separated in the region of Lake Chad. Today this does not represent a barrier, but from 10,000 until 7,000 years ago the much larger Mega-Chad lake stretched from the Sahara to the forest areas of west Africa. This lake was much larger than any lake today but has been drying progressively over the past 7,000 years due to changes in river courses that used to feed it and to global climate changes.

Much of the diversification of tortoise genera in the southern hemisphere coincides with the expansion of ice sheets in Antarctica and globally cooler climates.

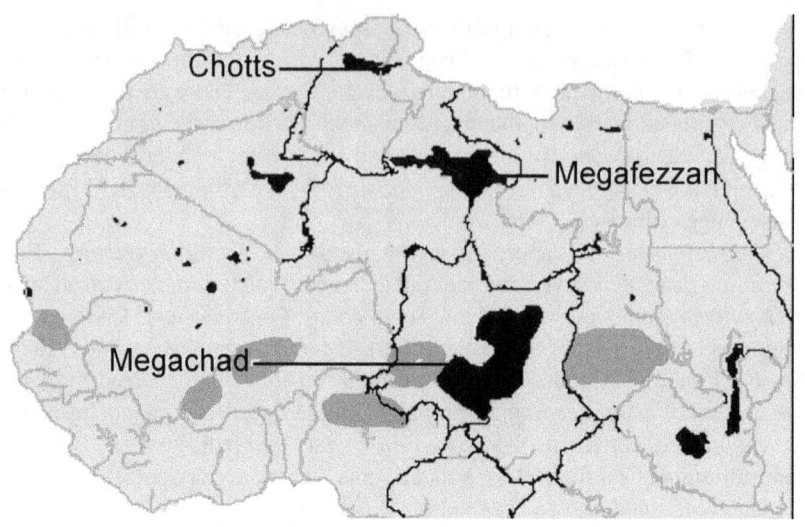

The great lakes of North Africa 10,000 years ago (black) and the range of the African spurred tortoise (grey) divided by Lake Megachad.

These would have resulted in reduced sea level and consequent exposure of land bridges and large island areas (since submerged). Movement of *Geochelone* out of North Africa, eastwards into Arabia and eventually as far as India would have been facilitated by lowered sea levels. The result of this movement was the two living Asian *Geochelone*: the Indian starred tortoise *G. elegans* and the Burmese starred tortoise *G. platynota*. As with their large north African ancestors they are found in dry forest and grassland areas. They have spread throughout dry

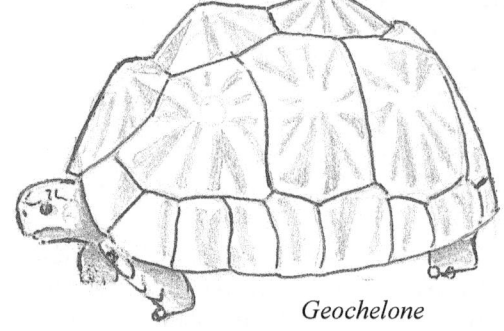

Geochelone

coastal areas of Pakistan, India, Burma and Sri Lanka. The two species are essentially geographic isolates of the same form, separated by the large marshy Baramaputra delta in Bangladesh which breaks up the dry habitats they favour, dividing the Indian starred in the west and the Burmese starred in the east. The Indian starred is restricted to two separate areas: one along the east coast of India and Sri Lanka, and another in north-west coastal India and Pakistan. These tortoises have characteristics of successful colonists, maturing relatively early (mating from 5-6 years) and nesting up to 4 times per year (with a total of around 24 eggs). The pyramiding of carapace scutes in the Indian starred resembles that often seen in captive tortoises suffering from growth problems as a result of inappropriate diets. This tends to results from excessively fast growth and this may be true of the Indian

180

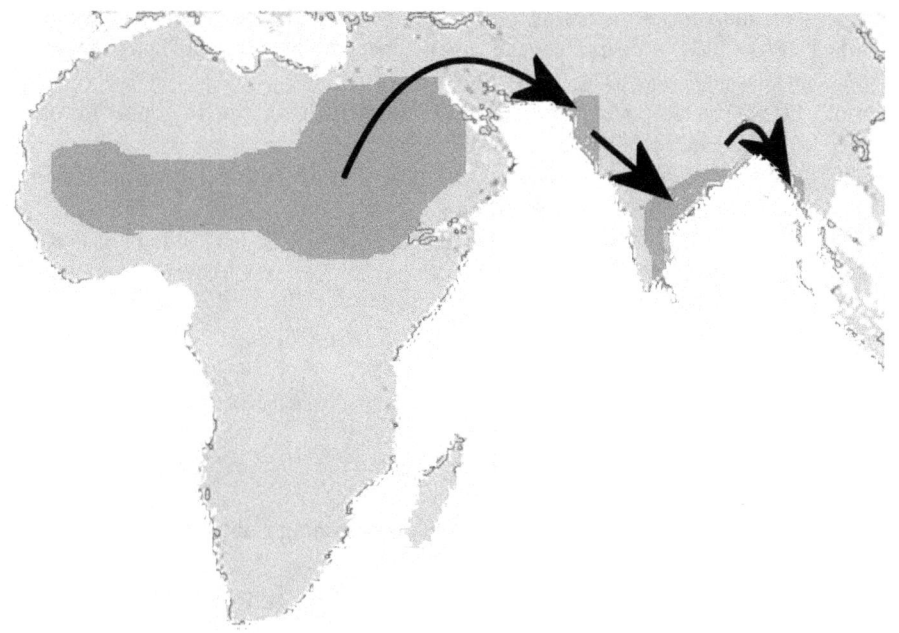

Distribution of *Geochelone* from 30 million years ago.

starred as well. Most have pyramidal scutes but some individuals are smooth; possibly the fast growth and early maturity of most individuals naturally leads to pyramiding. The Burmese resembles the Indian star but lacks such extreme pyramiding.

The attractive starred pattern in this species is very similar to that seen most obviously in the radiated tortoise *Astrochelys radiata* and the tent tortoises *Psammobates*. The close resemblance of these tortoises could be due to them being selected to have the same pattern which might be well camouflaged in long grass. However, not all of them are particularly strongly associated with grasslands and it does not seem a particularly good camouflage. It may be more probable that the beutiful pattern is really an accident of growth; strongly marked hatchlings may be well camouflaged but as they grow the spots on the scutes become drawn out into long rays, creating an obvious convergent pattern.

References

Auffenberg, W. 1963. The fossil testudinine turtles of Florida, genera *Geochelone* and *Floridemys*. *Bull. Fl. State Mus.* **7**(2):53-97

Auffenberg, W. 1974. Checklist of fossil land tortoises (Testudinidae). *Bull. Fl. State Mus., Biol. Sci.* **18**: 121-251

Auffenberg, W. 1976. The genus *Gopherus* (Testudinidae): Part 1. Osteology and relationships of extant species. *Bull. Fl. State Mus., Biol. Sci.* **20**(2): 47-110

Boycott, R.C. & O. Bourquin. 1988. *The South African Tortoise Book: A guide to South*

African tortoises, terrapins, and turtles. Southern Book Publ. Johannesburg

Bramble, D.M. 1974. Occurrence and significance of the Os transiliens in gopher tortoises. *Copeia* **1974**: 102-109

Bramble, D. M. 1982. *Scaptochelys*: generic revision and evolution of gopher tortoises. *Copeia*. **1982**(4): 852-867

Branch, W.R. 2007. A new species of tortoise of the genus *Homopus* (Chelonia: Testudinidae) from southern Namibia. *African Journ. Herpetol.* 56: 1-21.

Broadley, D.G. 1989. *Geochelone pardalis* Leopard Tortoise. In I.R. Swingland & M.W. Klemens (eds.) *The Conservation Biology of Tortoises.* IUCN/SSC Occ. Pap. No. 5, Gland.

Broadley, D.G. 1992. The savanna species of. *Kinixys*. (Testudinidae). *J. Herpetol. Assoc. Afr.* **40**: 12-13.

Broadley, D.G. 1993. Ankylosis in the shell bones of *Kinixys homeana* (Chelonii, Testudinidae). *Journ. Herpetol. Ass. Afr.* **42**

Claude, J. & H. Tong. 2004. Early Eocene testudinoid turtles from Saint Papoul, France. *Oryctos* **5**: 3-45

Crumly, C.R. 1982. A cladistic analysis of *Geochelone* using cranial osteology. *J. Herp.* **16** (3): 215-234

Crumly, C.R. 1984. *The evolution of the Testudinidae*. Unpublished Ph.D. thesis. Rutgers The State University

Daniels, S.R., M.D. Hofmeyr, B.T. Henen & E.H.W. Baard. 2010. Systematics and phylogeography of a threatened tortoise, the speckled padloper. *Anim. Conserv.* **13**: 237-246

Daniels, S.R., M.D. Hofmeyr, B.T. Henen & K.A. Crandall. 2007. Living with the genetic signature of Miocene induced change: Evidence from the phylogeographic structure of the endemic angulate tortoise *Chersina angulata. Mol. Phyl. Evol.* **45**: 915–926.

Danilov, I.G. & A.O. Averianov. 1997. New data on the turtles from the early Eocene of Kirghizia. *Russ. Journ. Herpetol.* **4**: 4045

Drake, N. & C. Bristow. 2006. Shorelines in the Sahara: geomorphological evidence for an enhanced monsoon from palaeolake Megachad. *The Holocene* **16**(6): 901-911

Fritz, U., S.R. Daniels, M.D. Hofmeyr, J. González, C.L. Barrio-Amorós, P. Siroký, A.K. Hundsdorfer & H. Stuckas. 2010. Mitochondrial phylogeography and subspecies of the wide-ranging sub-Saharan leopard tortoise *Stigmochelys pardalis* (Testudines: Testudinidae) – a case study for the pitfalls of pseudogenes and GenBank sequences. *J. Zool. Syst. Evol. Res.*

Gerlach, J. 2001. The '*Geochelone*' problem in tortoise taxonomy. *Phelsuma* **9**(A): 1-32

Harrison, T. 2011. Tortoises (Chelonii, Testudinidae). In: T. Harrison (ed.), *Paleontology and Geology of Laetoli: Human Evolution in Context. Volume 2: Fossil Hominins and the Associated Fauna*, Vertebrate Paleobiology and Paleoanthropology, DOI 10.1007/978-90-481-9962-4_17

Holroyd, P.A. & J.F. Parham. 2003. The antiquity of African tortoises. *Journ. Vertebr.*

Paleontol. **23**(3):688-690

Hutchison, J.H. 1996. Testudines. In: D. R. Prothero & R. J. Emry (eds.) *The Terrestrial Eocene-Oligocene transition in North America.* Cambridge University Press.

Hutchison, J.H. 1998. Turtles across the Paleocene/Eocene Epoch boundary in west-central North America. In: M.P. Aubry, S.G. Lucas & W.A. Berggren (eds) *Late Paleocene-Early Eocene Climatic and Biotic Events in the Marine and Terrestrial Records.* Columbia University Press, New York

Joyce, W.G., J.F. Parham & J.A. Gauthier. 2004. Developing a protocol for the conversion of rank-based taxon names to phylogenetically defined clade names, as exemplified by turtles. *Journ. Paleontol.* **78**(5): 989-1013

Kear, B.P. 2009. Evolution of gigantic tortoises from the Neogene of Europe. *Sylvester-Bradley Reports* **73**: 67

Kuyl, A.C. van der, D.L.P. Ballasina, J.T. Dekker, J. Maas, R.E. Willemsen & J. Goudsmit. 2002. Phylogenetic Relationships among the Species of the Genus *Testudo* (Testudines: Testudinidae) Inferred from Mitochondrial 12S rRNA Gene Sequences. *Mol. Phylogenet. Evol.* **22**(2): 174-183

Lapparent de Broin, F. 2000. African chelonians from the Jurassic to the present: phases of development and preliminary catalogue of the fossil record. *Paleontologia Africana* 36:43–82

Lapparent de Broin, F. 2003. *Neochelys* sp. (Chelonii, Erymnochelyinae), from Silveirinha, early Eocene, Portugal. *Ciências da Terra* **15**

Le, M., C.J. Raxworthy, W.P. McCord & L. Mertz. 2006. A molecular phylogeny of tortoises (Testudines: Testudinidae) based on mitochondrial and nuclear genes. *Mol. Phyl. Evol.* **40**(2): 517-531

Meylan, P.A. & W. Sterrer. 2000. *Hesperotestudo* (Testudines: Testudinidae) from the Pleistocene of Bermuda, with comments on the phylogenetic position of the genus. *Zool. J. Linn. Soc.* **128**: 51-76.

Morafka, D.J. & K.H. Berry. 2002. Is *Gopherus agassizii* a Desert-Adapted Tortoise, or an Exaptive Opportunist? Implications for Tortoise Conservation. *Chel. Conser. Biol.* **4**:263-287

Murphy, R.W., K.H. Berry, T. Edwards, A.E. Leviton, A. Lathrop & J.D. Riedle. 2011. The dazed and confused identity of Agassiz's land tortoise, *Gopherus agassizii* (Testudines, Testudinidae) with the description of a new species, and its consequences for conservation. *ZooKeys* **113**: 39-71

Palkovacs. E.P., J. Gerlach & A. Caccone. 2002. The evolutionary origin of Indian Ocean tortoises (*Dipsochelys*). *Mol. Phylogenet. Evol.* **24**(2): 216-227

Reynoso, V.-H. & M. Montellano-Ballesteros. 2004. A new giant turtle of the genus *Gopherus* (Chelonia: Testudinidae) from the Pleistocene of Tamaulipas, México, and a review of the phylogeny and biogeography of gopher tortoises. *Journ. Vert. Palaeont.* **24**: 822-837

Takahashi, A., H. Otsuka & R. Hirayama. 2003. A new species of *Manouria* (Testudines: Testudinidae) from the Upper Pleistocene of the Ryukyu Islands, Japan. *Palaeontol. Res.* **7**: 195-217

Takahashi, A., H. Otsuka & H. Ota. 2008. Systematic Review of Late Pleistocene turtles (Reptilia: Chelonii) from the Ryukyu Archipelago, Japan, with special reference to paleogeographical implications. *Pacific Science* **62**(3): 395-402.

Vlachos, E. 2011. *Contribution to the study of gigantic tortoises in stratigraphy and palaeogeography of the Neogene of Macedonia, Greece.* Unpublished MSc thesis, Aristotle University of Thessaloniki

Wall, W.P. & D. Maddox 1998. Reassessment of characteristics determining generic affinity in *Gopherus* and *Stylemys* (Testudinidae) from the White River Group, Badlands National Park. *National Park Service Paleontological Research* Vol. 3 Technical Report NPS/NRGRD/GRDTR-98/1 pp. 8-12

Winokur, R.M. & J.M. Legler. 1975. Chelonian mental glands. *Journ. Morphol.* **147**: 275-291

14. *Testudo* and its allies

The most familiar of all chelonians are the largely Mediterranean tortoises of the genus *Testudo*. This genus extends from North Africa and Southern Europe, through the Middle East as far east as Iran. There is a reasonable level of morphological diversity and some 20 species have been described. Some of these are sometimes separated into their own genera: *Agrionemys, Eurotestudo, Chersus* and *Pseudotestudo*. The main complexity lies in the *T. graeca/ibera* complex.

The earliest tortoises possibly related to *Testudo* may be the east Asian *Ergilemys insolitus* and "*E.*" *bruneti* from 55-23 million years ago but the relationships of these species are highly questionable and an origin closer to 15 million years ago is more plausible. More recent fossils have been referred to the living genus *Agrionemys* from west Asia. The Mediterranean tortoises are also closely related to the African pancake tortoise *Malacochersus tornieri* and to the Asian forest tortoises *Indotestudo*.

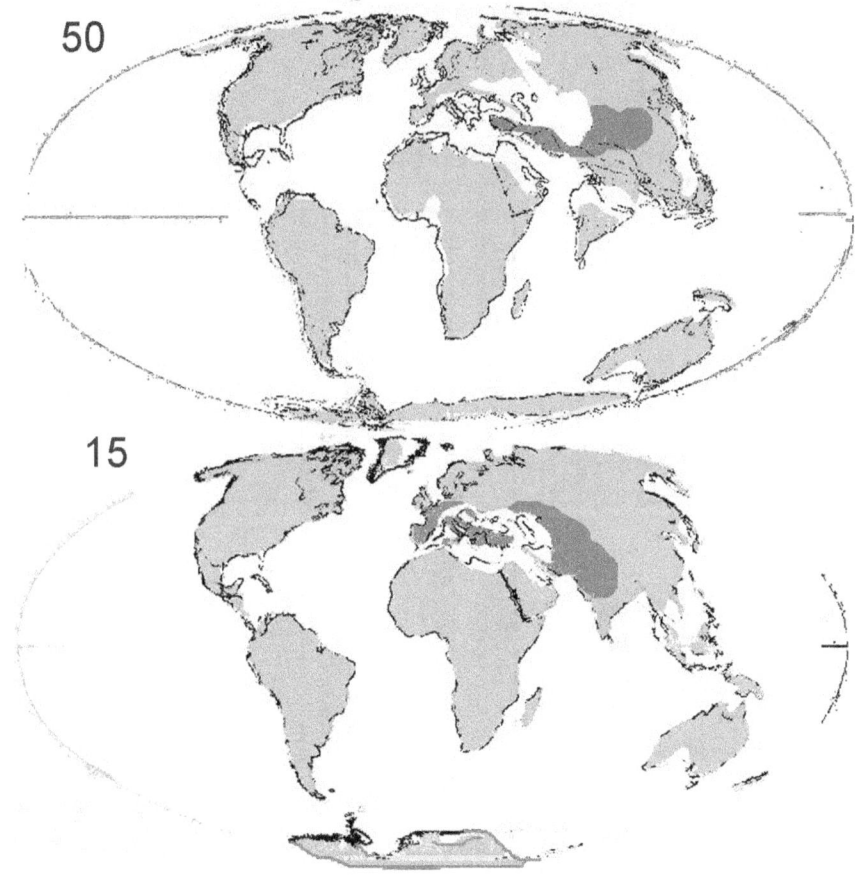

Distribution of early *Testudo* relatives from 50 million years ago.

These 'testudonans' probably evolved at least 15 million years ago in Asia although the first definite fossils are North African ones from 3 million years ago. Between 10 and 6 million years ago these tortoises were able to disperse throughout Europe, Asia and into east Africa through land connections formed due to the collision of Arabia into Turkey, creating the land bridge that today connects Europe, Africa and Asia.

Molecular studies of the 'testudonans' (*Testudo* in its widest sense, along with *Indotestudo* and *Malacochersus*) are highly contradictory, suggesting that *Malacochersus* may be related to *Indotestudo* or *Agrionemys* and that *Eurotestudo* may group with *Testudo* or, more often, with *Agrionemys*. Although *Malacochersus* is the strangest-looking of all tortoises with its very flat shell, its skeleton is very similar to *Agrionemys*; but then both are at least partially burrowing or crevice dwelling species so may have converged morphologically. Both genera are arid environment species, living in rock crevices in the case of the pancake tortoise and burrowing in the case of *Agrionemys*. *Agrionemys*, the steppe tortoise, differs from the Mediterranean *Testudo* species in several respects, most obviously the presence of four instead of five claws on the front feet. These different genera diverged around 26 million years ago (according to DNA calculations), dividing into distinct geographical groups: south and east Asian (*Indotestudo*), west Asian (*Agrionemys*), Mediterranean (*Testudo/Eurotestudo*), and east African (*Malacochersus*).

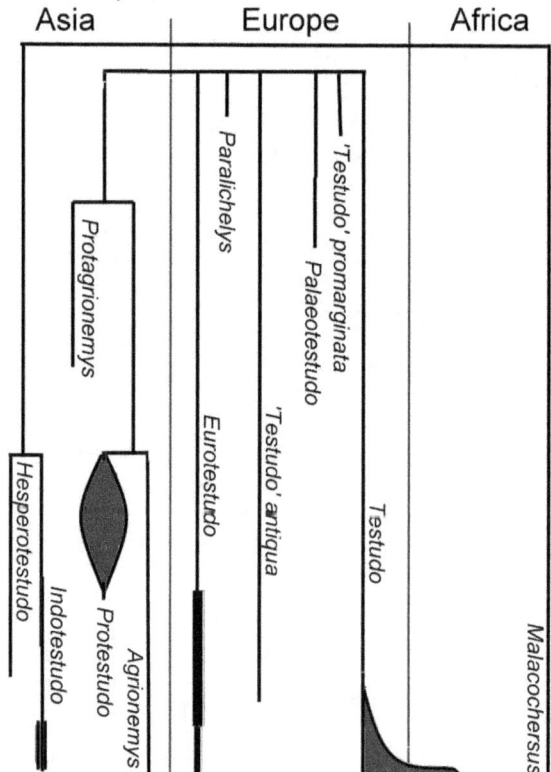

Phylogeny of *Testudo* and relatives.
186

The ancestral *Indotestudo* probably ranged from the Western Ghats of India to Indonesia. Drying climates during the Ice Ages fragmented the forest habitats and isolated the three species alive today. These are restricted to three geographically separate areas: a western species (Travancore tortoise *I. travancorica* in western India), a central species (the red-nosed *I. elongata* in eastern India to south-east Asia and southern China) and an isolated eastern species (Sulawesi tortoise *I. forstenii* on Sulawesi island). These are found in dry forests in the west and wetter forests in the centre and east. They largely feed on fruit and reproduce relatively slowly, laying only 4-5 eggs at a time, habits which would restrict their ability to disperse between forest patches. These tortoises are unusual in showing conspicuous colour change during breeding: in the breeding season males develop distinctively white heads. This sort of change is seen in some terrestrial geoemydids. *Indotestudo* tortoises have uncertain relationships; some studies suggest that the Travancore is the sister species to the red-nosed, but a relationship between the Travancore and the Sulawesi may also be possible, although this makes little geographical sense. This confusion is probably due to the three forms becoming isolated at more or less the same time, so forming three equally related populations. The genus has an obscure history; the only possible fossil is *'Hesperotestudo' kalganensis* from China within the last 5 million years. Other Chinese species from the time may be closer to true *Testudo*, such as *"Testudo" hipparionum*.

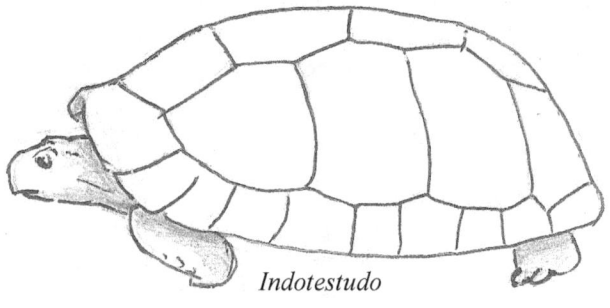

Indotestudo

Malacochersus comprises a single species, the pancake tortoise *M. tornieri*. This is the strangest and most specialised of all the living tortoises. It is restricted to rocky outcrops in the savannah of Kenya and Tanzania. During cool times of day it forages over the outcrops for herbs, fruit and invertebrates, retreating into rock crevices as temperatures rise. The shell of this species is distinctively flattened and flexible due to the reduction in ossification of the shell bones. This bone reduction was originally interpreted to be evidence of bone disease in the early specimens, but in fact it is a normal change in development. This essentially juvenile soft shell allows the tortoises to squeeze themselves into tight crevices in the rocks to avoid predators. At one time it was thought that these tortoises could inflate themselves to wedge into crevices, but this is now known not to be the case. With its restriction to dry rocky outcrops there is no useful fossil record for the species. Until 8 million years ago most of East Africa was probably forested; geological and climatic changes resulted in the development of the pancake tortoise favourite habitats from around 7 million years ago. The pancake tortoise probably evolved in arid areas further north, spreading into its present range as the east African savannahs formed. The ancestral northern populations would have been lost to increasing aridification as the Sahara desert formed 3.5 million years ago.

187

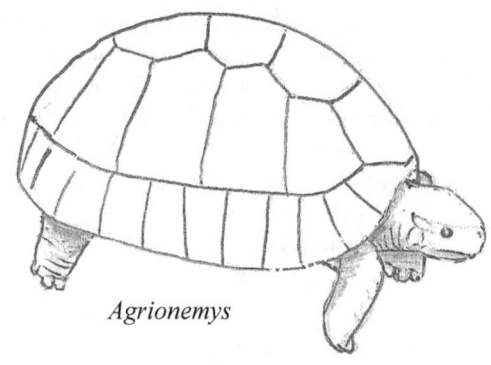

Agrionemys

All possible fossil ancestors of *Agrionemys* are found in an area between Kazakhstan and Iran, within the range of the living species. The earliest possible ancestor of *Agrionemys* appears to be *Protagrionemys turgaica* from Kazakhstan 15 million years ago. *Agrionemys* itself is known from at least 9 million years ago, the early forms being ascribed to "*Protestudo*", which first appears with *P. bessarabica* of Iran and Afghanistan. At least 16 fossil *Agrionemys/Protestudo* species have been described. Three living species (or subspecies) exist, representing geographical variants: Horsfield's tortoise *T. horsfieldi* of northern Iran to Afghanistan and Pakistan, the Kazakhstan steppe tortoise *T. kazachstanica* from Kazakhstan to western China, and the Kopet-Dag steppe tortoise *T. rustamovi* of the eastern coast of the Caspian Sea. Two other genetically distinct forms have been recognised but are not often referred to: the Fergana valley steppe tortoise *T. bogdanovi* of Kyrgyzstan, Tajikistan, Turkmenistan, Uzbekistan and the Turkmenistan steppe tortoise *T. kuznetzovi* of Turkmenistan and Uzbekistan. These groups are only partially in agreement with mitochondrial DNA evidence which identifies three groups.

The main genetic group of *Agrionemys* is spread throughout the northern range of the genus, to the southern end of the Caspian Sea. This corresponds to the Kazakhstan steppe tortoise (*kazakhstanica*, including *rustamovi*). A few of this group also occur in central Iran in the range of what are normally thought to be Horsfield's. True Horsfield's (*horsfieldi*, including *baluchiorum*) make up the second most widespread group, found in the south-east, mainly in Iran and Afghanistan. The south-eastern shores of the Caspian Sea are shared by both of these groups (including the range of the Kopet-Dag tortoise) which seems to be an area of overlap between them. There is an isolated un-named genetic form in the Fergana Valley, supposedly in the range of the Kazakhstan steppe, but apparently more closely related to Horsfield's. This latter form also spreads up the eastern part of the range. It seems that during the Ice Ages this species was isolated in three main areas (north, south-east and the Fergana Valley); they have since expanded from their refugia, and overlapped in the region of the Caspian Sea and the east. At present it is not known if the nuclear DNA supports the present subspecies patterns, suggesting some hybridisation in the areas of overlap, or whether the morphologically based subspecies are completely misleading. The Mongolian steppe tortoise *T. terbishi* is present in Mongolia but may be an introduced population of one of the other forms; its genetic identity is not known.

These are specialised arid environment tortoises, with highly developed burrowing abilities; their front feet are broad and powerful, with short toes, enabling them to dig deep burrows in which they hibernate in cold climates or aestivate in hot

188

climates. In southern Uzbekistan they may be active for as little as three months of the year. They are hardy animals as their harsh environment requires. This has contributed to them becoming the commonest tortoise species in trade in Europe in the past few decades, with an estimate of some 641,700 animals being exported from Uzbekistan between 1980 and 1999 and over 160,000 between 1995-1999 being exported by Kazakhstan, Uzbekistan, Tajikistan and Russia. Although many of these animals are imported as 'captive bred' most probably originate from eggs obtained from wild captured females.

Eurotestudo originated around 12 million years ago with several fossil species from Spain (*E. globosa* and *E. lunellensis*), France (*E. pyrenaica*), Italy (*E. globosa*) and Poland (*E. szalai*). Of the living species Hermann's tortoise *E. hermanni* may have evolved within the last 5 million years, with the first definite fossils of this species

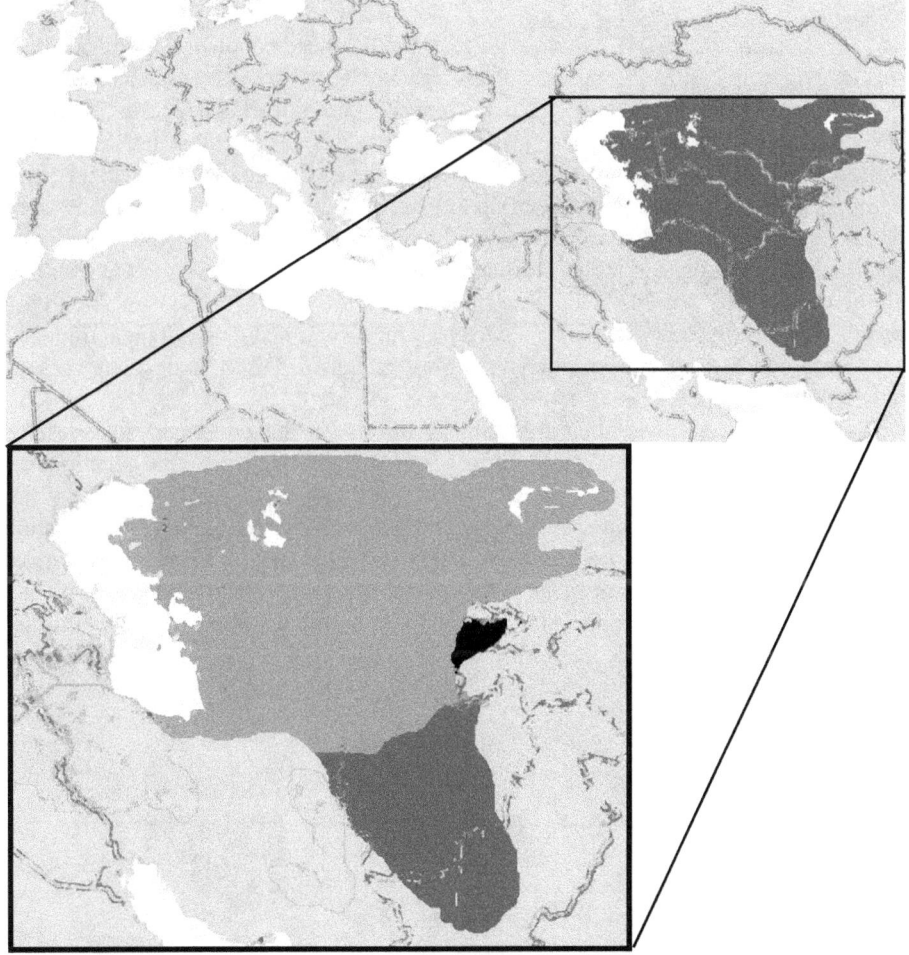

Distribution of the *Agrionemys* genetic groups: Kazakhstan steppe tortoise (dark grey), Horsfield's tortoise (light grey), Fergana valley steppe tortoise (black).

from Corsica 1.5 million years ago. The living *Eurotestudo* are variously classified as species or subspecies. Boettger's *E. boettgeri* is a sister species or subspecies of Hermann's tortoise; it could be further divided with *hercegovinensis* of the Adriatic being distinctive (usually lacking an inguinal scute). Boettger's is found throughout Greece, northwards along the Adriatic coast. Hermann's is found in coastal areas of the western Mediterranean. This differs from Boettger's in having a divided supracaudal and in colouration. These three have clearly evolved in geographical isolation (allopatrically), differing in minor but recognisable characteristics and having non-overlapping ranges. This probably represents isolation in Italy and Greece during glacial times for Hermann's and Boettger's respectively. *E. hercegovinensis* is probably a recent, only partially separated geographical variant. These animals were being exploited by humans as far back as 30,000 years ago as shown by butchered tortoise bones from Neanderthal sites in Spain.

Mitochondrial DNA studies suggest that all these populations form a single species which is still in the process of divergence. There are distinct genetic groups in the western Mediterranean, Southern Peloponnese, Bulgaria-Greece and Croatia-Greece. The western Mediterranean group are Hermann's tortoise, these are genetically relatively uniform despite having been present in the area, including the Corsican and Sardinian islands, for the past 2.5 million years. The western-most populations were driven to extinction at the end of the last Ice Age when extremely cold conditions persisted from 39,500 to 38,000 years ago. The low genetic diversity is the result of all the present day tortoises being descended from a few survivors. Some of the living populations also descend from human introduction, such as in the case of the introduced Mallorcan and Menorcan populations. On these islands tortoise remains date back to only 3,000 years ago.

There is more variation in Boettger's tortoise as the landscape of the eastern Mediterranean provided more warm refuges from the cold Ice Ages. One genetic form is isolated on the west coast of the Balkans (the locality for *hercegovinensis*) by the Dinarid and Pindos mountain chains, as is a group on the western slope of the Taygetos Mountains in the southern Peloponnese. These tortoises lack inguinal scutes (unlike all other *Testudo-Eurotestudo* species) due to the expansion of the abdominal scutes. In addition some of these animals have a highly unusual form of the inguinal buttress, which forks at the top, inside the shell.

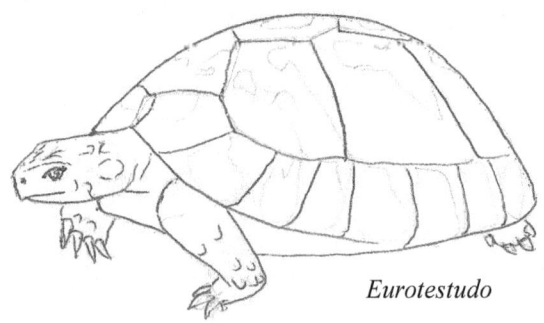

Eurotestudo

Mountain chains acting as barriers should mean that the Iberian population, isolated by the Pyrenees, should be genetically different; however there is very little difference between Spanish and French populations, suggesting only very recent isolation. Western Mediterranean populations group as Spain, France-Italy, Sicily-Corsica-Sardinia (introduced to the Balearics and to the Ebro delta in Spain), with isolated forms in parts of Italy, Corsica and Sardinia. It is possible that the island tortoises are now all introduced although there were originally totally separate forms on the three naturally occupied islands. Today there is a mixture on all except Sicily, but the differences are all very minor. The Apennines should separate Italian populations, but these mountains are relatively low and there do not seem to be any special Italian forms.

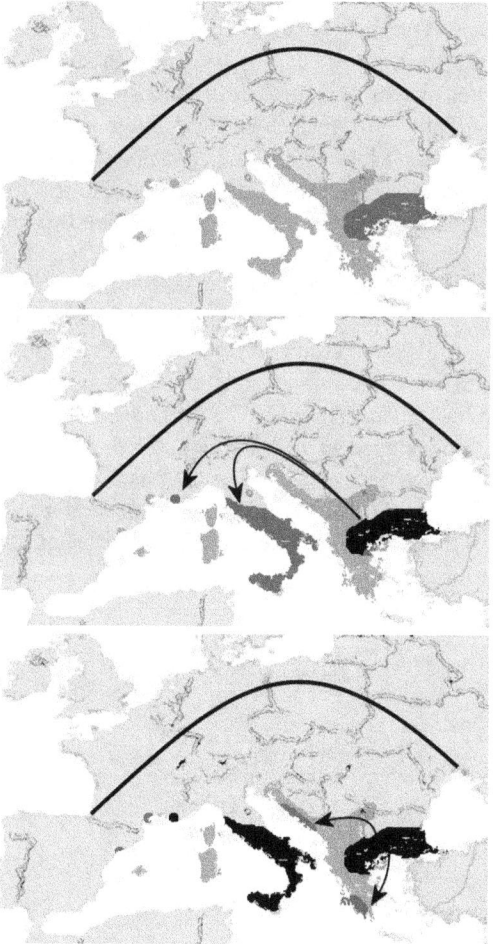

Distribution of *Eurotestudo* from 12 million years ago.

black line - northern extent of fossils from 12 million years ago.
light grey - current range
dark grey - probable range during the Ice Ages.

Expansion over the last 40,000 years.

dark grey - first Hermann's tortoise populations.

Later expansion.

dark grey - western Balkans and Peleponnese populations, other areas colonised more recently.

Fossils from 34-5 million years ago are of uncertain identification. There are several fossil western *Testudo*-like species without a plastral hinge (i.e. similar to *Eurotestudo*) which can be divided into the *T. promarginata, Palaeotestudo* and *"T." antiqua* groups. The *T. promarginata* group (20 million years ago in France and Germany) have a straight epiplastral lip similar to the living *Testudo graeca 'antakyensis'* (a form of *T. g. terrestris*). *Palaeotestudo* includes *P. canetotiana* (France 12-13 million years ago), *P. mellingi* (Austria and France 17 million years ago) and *P. angustihyoplastralis* (20 million years ago of Austria), and possibly *"Testudo" catalaunica* (France 10 million years ago), *"Testudo" semenensis* (Tunisia, 5 million years ago) and *Paralichelys catalaunicus* (Spain, 20 million years ago). These differ from *T. promarginata* in their abrupt shell margins and convex posterior of the shell; these also distinguish the genus from *Eurotestudo*. *"T." antiqua* group is similar to *Agrionemys* in some respects. Species referred to this group include the 23-5 million years ago species *Testudo escheri* (Germany) and *'?Cheirogaster' arrahonensis* (Spain). These early species may be ancestral to *Testudo, Eurotestudo* and *Agrionemys*.

True *Testudo* probably started to diverge from a *Eurotestudo*-like ancestor some 15 million years ago. This would probably have resembled *T. promarginata*, starting to develop a plastral hinge and a long plastral lobe like a true *Testudo*. This would have been similar to *'Paleotestudo' canetotiana* 7 million years ago, and the first true first *Testudo* to appear, *T. marmorum* in Greece. The genus appears to have originated in eastern Europe and spread west 6 million years ago when *T. amiatae* is found in Italy. The westward movement would have been facilitated by the Messanian salinity crisis when the Mediterranean dried out from 6 million years ago until the Zanclean flood of 5.3 million years ago, enabling easy movement around the edges of a great salt lake.

T. marmorum is similar to *T. marginata* in that both have a hinge in the posterior lobe of the plastron, typical of true *Testudo*. The hinge between the hypo- and xiphiplastra is particularly well developed in *T. marginata* and *T. kleinmanni* but quite limited in other living *Testudo*. Early fossils showing the characteristic *Testudo* plastron include *T. bosborica, T. bulcarica, T. eldarica, T. kalganesis, T. kenitrensis, T. pecorinii* and *T. semenensis*. Most of these are from the eastern Mediterranean, although *T. pecorinii* is from Sardinia 3.5 million years ago.

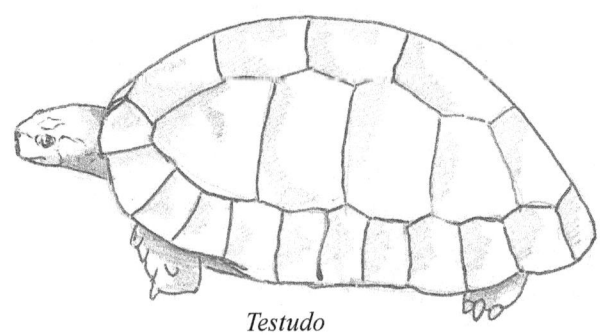

Testudo

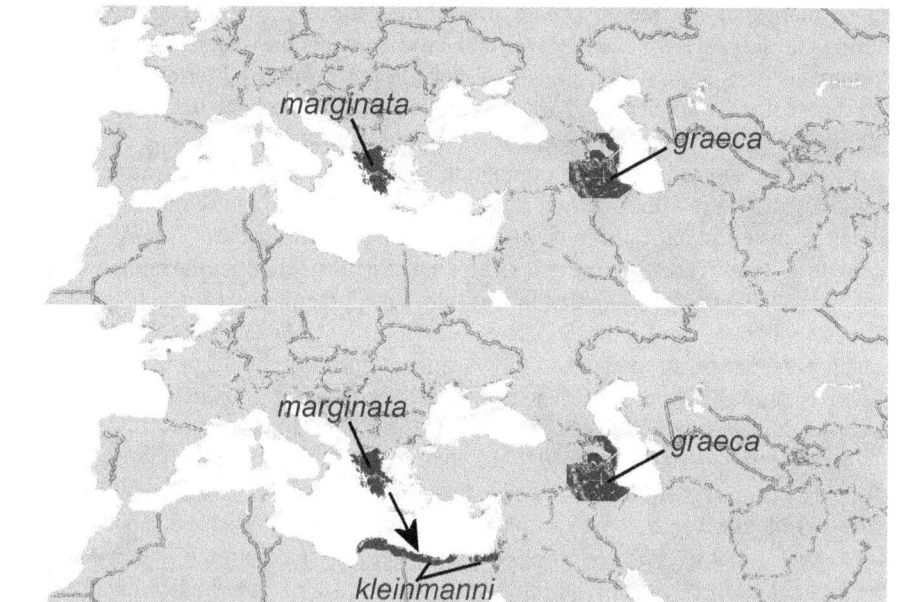

Origins of living *Testudo* species - *marginata* and the early range of *graeca* shown in dark grey (top), followed by the origin of *kleinmanni* by dispersal from Greece.

Of the conventional *Testudo* species, the most distinctive is the marginated tortoise *T. marginata*. This forms a group ('*Chersus*') with Kleinmann's tortoise *T. kleinmanni*. The marginated tortoise is found in south-eastern Europe, whilst Kleinmann's is restricted to north Africa. These distributions cover the north-east and south-east coasts of the Mediterranean, a distribution that may have arisen from the extinction of intermediate eastern populations, isolation of the two populations by the flooding of the Mediterranean basin 5.3 million years ago or movement between the two areas. It has been suggested that this latter possibility is the most likely, although there is little evidence of transoceanic movement of *Testudo*. The oldest fossils from the range of Kleinmann's are no more than 3 million years old, suggesting that the present day distribution is not the result of isolation caused by flooding of the Mediterranean.

The marginated is a large, distinctive species with the exception of the Peloponnese population described as *T. weissingeri*. This seems to be a dwarf form of *marginata* (being under 22 cm long, whereas typical marginateds are over 26 cm) but resembles the fossil *T. marmorum*. *T. weissingeri* is also slightly different in shape, having relatively wide shells and larger shell openings. Males have broad plastra to facilitate mating with the broad females. Despite their morphological differences genetic studies have failed to separate *weissingeri* and *marginata* using mitochondrial DNA and nuclear DNA fingerprinting. It would seem that this form is questionable and it has been suggested that *weissingeri* is dwarfed due to local dietary factors, but they do apparently produce equally dwarfed offspring in captivity.

The populations of marginated tortoises in the west coast of the Balkans are found alongside Boettger's tortoise. The distinct form of Boettger's tortoise isolated during the Ice Ages by the Dinarid and Pindos mountains chains and the Taygetos Mountains does not have a comparable form in the marginated tortoise. It seems that the mountains do not act as a barrier in the marginated, probably because this species occurs up to 1300 m whereas Boettger's is mainly a lowland species.

The isolated Sardinian population is also somewhat different from the Greek marginateds. These smooth animals lack the marginal flaring and serration typical of the Greek tortoises and have been described as a separate subspecies: *T. marginata sarda*. As with *T. weissingeri* they are genetically indistinguishable from the marginated and it seems probable that this population was introduced, the unusual shape being inherited from a small number of introduced ancestors (a strong 'founder effect'). It is not known whether the Sardinian marginated tortoises were introduced in historical or prehistoric times. A naturally occurring island population did exist in the past with *T. marginata cretensis* being described from Crete around 2 million years ago.

Kleinmann's is restricted to the coast of Libya, and formerly Egypt. It is a very distinctive small, pale *Testudo* species. Eastern Egyptian and Israeli populations were formerly included in this species but are sometimes considered a distinct species: *T. werneri*. *T. werneri* is a small, largely pale species found from the Sinai peninsula to the Negev desert of Israel. The limited genetic studies of Israeli and Libyan tortoises indicate that these two forms are probably identical; all samples of Kleinmann's are found within the variation of the slightly more variable Werner's.

The most familiar and most complicated of all *Testudo* are undoubtedly the *T. graeca* complex. This covers the spur-thighed tortoises *T. graeca* and several forms regarded as distinct species, subspecies or local variants. Within this group at least 20 forms have been described and there is no consensus on how many are valid and whether they are species, subspecies or local races. Almost every confusion has been heaped on these animals; species have been described based on animals of unknown origin, variation has rarely been studied, names have been misapplied, animals have been moved around. Confusion even surrounds the naming of the type form, *T. graeca*; this sounds like it should be a Greek tortoise but is in fact a North African species named by Linnaeus in reference to the shell pattern and colouration which brought to his mind an ancient Greek vase.

Mitochondrial DNA and DNA fingerprinting identifies six groups within this complex. Four of these are found in the Caucasus: A (East Caucasus), B (North Africa and Mediterranean islands), C (western Asia Minor, south-eastern Balkans to the west-central Caucasus) and D (south and eastern Asia Minor and Levantine). Two other groups are restricted to Iran: E (northwest – central Iran) and F (eastern Iran). The east Caucasus and North African A and B are sister groups, as are the Balkans and Levantine (C and D). These groups are not apparent in nuclear fingerprints. Some of the morphologically distinctive forms (e.g. *T. g. armeniaca* and *T. g. floweri*) belong to groups containing morphologically very different forms (group A: *T. g. armeniaca*, *T. g. ibera*, *T. g. pallasi*; D: *T. g. anamurensis*, *T. g. antakyensis*, *T. g. floweri*, *T. g. ibera*, *T. g. terrestris*) in geographical but not morphological patterns. Conversely *T. g. ibera* tortoises are morphologically conservative but are spread over the genetic groups A,

C and D. This has been interpreted as being the result of morphological adaptation to environmental conditions. The genetic data suggest that about 10 subspecies exist and that these diverged about 1 million years ago. The genetic evidence points to the spur-thighed tortoise originating in the Caucasus (where it is still more diverse genetically than elsewhere). This region has the oldest fossils similar to spur-thighs: *T. burschaki* and *T. eldarica* (Georgia and Azerbaijan 10-8 million years ago). It seems probable that the living species had evolved in this area by about 4 million years ago.

The subspecies in the Caucasus today is *T. g. armeniaca*. This seems to have spread west and eastwards giving rise to Israel-Bulgaria and Iranian forms respectively around 4 million years ago. A related form also spread south and then westwards into North Africa around 2 million years ago (as indicated by North African fossils 2.5-2 million years ago) as a result of dispersal across the land-bridge that emerged when Africa collided with Turkey and Arabia 19-12 million years ago. *T. g. armeniaca* is now found mainly in Armenia, on the south-western shores of the Caspian Sea. It resembles Horsfield's tortoise and is also a burrowing form.

In North Africa *nabeulensis* is found in Tunisia and Algeria, *T. g. graeca-morokkensis* from Algeria to northern Morocco, *cyrenaica* in Libya (Cyrenaica peninsula) and *soussensis* in Morocco north of the Atlas Mountains and in the Souss Valley. Within these there is a division between *T. g. graeca* from eastern Morocco to Algeria, and *T. g. marokkensis* from north-western Morocco populations. In western Morocco two subspecies have been described: *T. g. lamberti* and

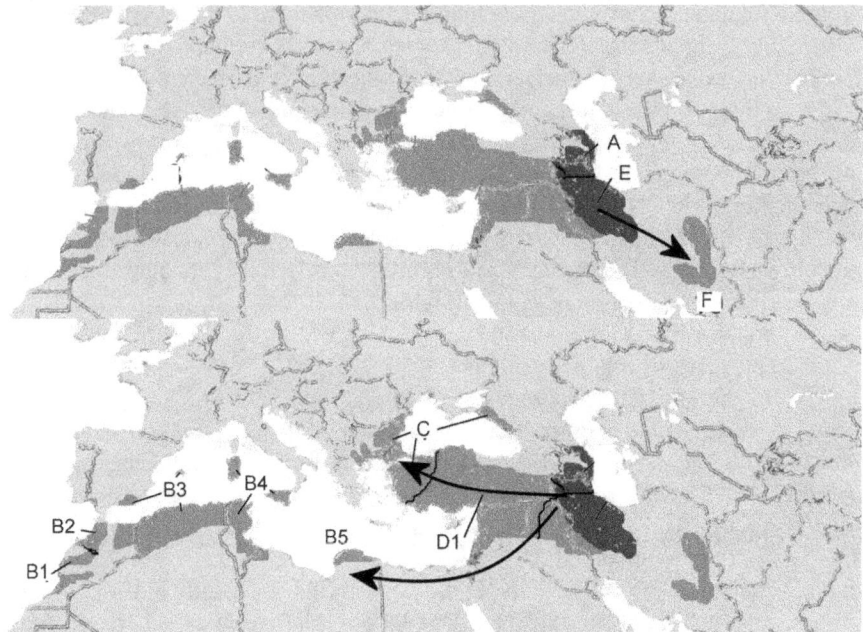

Spur-thighed tortoise origins, early range dark grey, present range light grey.
A - *armeniaca*; B1 - *soussensis*; B2 - *marokkensis*; B3 - *graeca*; B4 - *nabeulensis*; B5 - *cyrenaica*; C - *ibera*; D - *terrestris*; E - *perses*; F - *zarudnyi*

195

T. g. marokkensis. T. g. lamberti has more prominent spurs on its thighs and *T. g. marokkensis* has a flatter carapace but they seem to be identical genetically. It was thought that they were isolated to the north and south of the Rif Mountains, but it is now known that they range throughout these mountains and so should be considered a single form - *T. g. marokkensis*. Tortoises from the *T. g. nabeulensis* group have been introduced to Sicily and Sardinia, and one individual was found within the *T. g. graeca* population in east Algeria. The dates of these introductions are not known; Sardinia has been occupied by humans since at least prehistoric times and has been a major trader with ports located in the range of *T. g. nabeulensis*. It is possible that the tortoises were introduced by Phoenician traders several thousand years ago. Spain has a population of Moroccan spur-thighed tortoises, which also seem to be relatively recent arrivals, either by prehistoric introduction or by natural dispersal over water. Some *graeca* individuals also occur in the range of *T. g. soussensis*, either as a result of introduction or through movement along the Mediterranean coast or up the Moulouya River Valley. *T. g. graeca* genotypes in Libya are probably introduced. These were all divided somewhere between 1 and 1.4 million years ago in association with periods of increasing aridity fragmenting populations and leading to local genetic divergence. The division in Morocco between *T. g. graeca* and *T. g. marokkensis* is slightly more recent at around 700,000 years ago. North African tortoises inhabit a narrow range of habitats and morphological differences are largely the result of genetic diversification, not complex local environmental effects, unlike the eastern populations where arid-habitats promote small, pale forms and more humid environment favour larger, darker tortoises. A particularly interesting form is *T. g. 'whitei'*, this large, dark tortoise is less well marked than usual spur-thighed tortoises and much larger (20-30 cm long, compared to 15-20 cm for most other subspecies). Studies of this species have been plagued by a lack of information on the origin of supposed specimens. Its type locality was near Algiers and large Algerian and Moroccan tortoises have been assumed to belong to this form, with smaller ones being *T. g. graeca*. Samples from the type locality (so more reliably *whitei*) are indistinguishable from *T. g. graeca*. Its morphological distinctiveness is probably an ecological influence resulting from locally relatively humid valleys. This was briefly put into a different genus, *Furculachelys*, along with *T. g. nabulensis* on the basis of both forms having a forked suprapygal; however the shape of the suprapygals is less stable that was originally thought.

In the Eurasian subspecies of *T. graeca* there are two groups: the western (Israel to Bulgaria) and eastern (Iran). The western group has a southern subgroup from Iran to south and eastern Turkey (comprising the genetically indistinguishable *T. g. anamurensis, T. g. antakaiensis, T. g. terrestris* and *T. g. floweri*) and a northern group from north-west Turkey through Russia, Georgia and Bulgaria (being *T. g. ibera* and *T. g. nikolski*); these diverged 1.8 million years ago. The Iranian tortoises fall into two groups: eastern (*T. g. zarudnyi*) and north-western (the genetically indistinguishable *T. g. perses* and *T. g. buxtoni*) which diverged 3.6 million years ago. *T. g. buxtoni* been described from a small range on the southern shore of the Caspian Sea. *T. g. zarundryi* is an isolated species restricted to Iran; it is large and elongate for a *Testudo* species. In the northern group *T. g. nikolski* is restricted to the north-eastern coast of the Black Sea but should be included in *T. g. ibera*. This form is distinctive in appearance, with a rounded shell. Within the southern group *T. g. terrestris* is a moderately large, pale species found in Mesopotamia from the Turkey-Syria border to northern Iraq and south to Israel. *T. g. antakyensis* is unusual in having a straight or slightly curved epiplastral lip (as in *"T." promarginata*). This may be a juvenile reature retained by adults and caused by a small change in genes regulating development timing (paedomorphosis). This is a common phenomenon in many animal species, resulting in rapid morphological change with little genetic change; a clearer example is seen in the dwarf tortoises of the Indian Ocean which are described in the next chapter.

References

Abbazzi, L., S. Carboni, M. Delfino, G. Gallai, L. Lecca & L. Rook. 2008. Fossil vertebrates (Mammalia and Reptilia) from Capo Mannu Formation (Late Pliocene, Sardinia, Italy), with description of a new *Testudo* (Chelonii, Testudinidae) species. *Rev. Ital. Palaeontol. Stratigr.* **114**: 119-132

Cerling, T.E., J.G. Wynn, S.l A. Andanje, M.I. Bird, D.K. Korir, N.E. Levin, W. Mace, A.N. Macharia, J. Quade & C.H. Remien. 2011. Woody cover and hominin environments in the past 6 million years. *Nature* **476**(7358): 51

Chesi, F., M. Delfino & L. Rook. 2009. Late Miocene *Mauremys* (Testudines, Geoemydidae) from Tuscany (Italy): Evidence of Terrapin Persistence after a Mammal Turnover. *Journ. Paleontol.* **83**: 379-388

Chkhikvadze, V.M. 2001. On the systematic position of some Asian fossil turtles. *Works of the Tbilisi State Pedagogical University named after S.-S.Orbeliani* **10**

Fèlix, J., J. Budó, X. Capalleras & R. Mascort. 2004. The fossil register of the genera *Testudo, Emys* and *Mauremys* of the Quaternary in Catalonia. *Chelonii* **4**

Fritz, U., M. Auer, A. Bertolero, M. Cheylan, T. Fattizzo, A.K. Hundsdörfer, M. Martín Sampayo, J.L. Pretus, P. Siroky & M. Wink. 2006. A rangewide phylogeography of Hermann's tortoise, *Testudo hermanni* (Reptilia: Testudines: Testudinidae): implications for taxonomy. *Zool. Script.*

Fritz, U., M. Auer, M.A. Chirikova, T.N. Duysebayeva, V.K. Eremchenko, H.G. Kami, R.D. Kashkarov, R. Masroor, Y. Moodley, A. Pindrani, P. Široký & A.K. Hundsdörfer. 2009. Mitochondrial diversity of the widespread Central Asian steppe tortoise (*Testudo horsfieldii* Gray, 1844): implications for taxonomy and relocation of confiscated tortoises. *Amphibia-Reptilia* **30**: 245-257

Fritz, U., D.J. Harris, S. Fahd, R. Rouag, E.G. Martínez, A. Giménez Casalduero, P. Široký, M. Kalboussi, T.B. Jdeidi & A.K. Hundsdörfer. 2009. Mitochondrial phylogeography of *Testudo graeca* in the Western Mediterranean: Old complex divergence in North Africa and recent arrival in Europe. *Amphibia-Reptilia* **30**: 63-80

Fritz, U., P. Kiroký, H. Kami & M. Wink. 2005. Environmentally caused dwarfism or a valid species - Is *Testudo weissingeri* Bour, 1996 a distinct evolutionary lineage? New evidence from mitochondrial and nuclear genomic markers. *Mol. Phyl. Evol.* **37**(2): 389-401

Fritz, U., A.K. Hundsdörfer, P. Široký, M. Auer1, H. Kami, J. Lehmann, L.F. Mazanaeva,

O. Türkozan & M. Wink. 2007. Phenotypic plasticity leads to incongruence between morphology-based taxonomy and genetic differentiation in western Palaearctic tortoises (*Testudo graeca* complex; Testudines, Testudinidae). *Amphibia-Reptilia* **28**: 97-121

Iverson, J.B., P.Q. Spinks, H.B. Shaffer, W.P. McCord & I. Das 2001. Phylogenetic relationships among the Asian tortoises of the genus *Indotestudo* (Reptilia: Testudines: Testudinidae). *Hamadryad* **26**(2):271-274

Kuyl, A.C van der, D.L.P. Ballasina & F. Zorgdrager. 2005. Mitochondrial haplotype diversity in the tortoise species *Testudo graeca* from North Africa and the Middle East. *BMC Evol. Biol.* **5**: 1-29

Lapparent de Broin, F. de, R. Bour, J.F. Parham & J. Perälä. 2006. *Eurotestudo*, a new genus for the species *Testudo hermanni* Gmelin, 1789 (Chelonii, Testudinidae). *Comp. Rend. Palevol.* **5**(6): 803-811

Le, M., C.J. Raxworthy, W.P. McCord & L. Mertz. 2006. A molecular phylogeny of tortoises (Testudines: Testudinidae) based on mitochondrial and nuclear genes. *Mol. Phyl. Evol.* **40**: 517-531

Parham, J.F., J.R. Macey, T.J. Papenfuss, C.R. Feldman, O. Türkozan, R. Polymeni & J. Boore. 2006. The phylogeny of Mediterranean tortoises and their close relatives based on complete mitochondrial genome sequences from museum specimens. *Mol. Phyl. Evol.* **38**: 50–64

Parham, J.F., O. Türkozan, B.L. Stuart, M. Arakelyan, S. Shafei, J. R. Macey, Y.L. Werner & T.J. Papenfuss. 2006. Genetic Evidence for Premature Taxonomic Inflation in Middle Eastern Tortoises. *PCAS* **57**: 955-964

Perala, J. 2001. Biodiversity in relatively neglected taxa of *Testudo* L., 1758 S.L. *Chelonii* **3**

Spinks, P.Q., H.B. Shaffer, J.B. Iverson & W.P. McCord. 2004 Phylogenetic hypotheses for the turtle family Geoemydidae. *Mol. Phyl. Evol.* **32**: 164-182

Široký, P. & U. Fritz. 2007. Is *Testudo werneri* a distinct species? *Biologia, Bratislava* **62**(2): 1-4

Theile, S. 2002. Ranching and breeding of *Testudo horsfieldi* in Uzbekistan. *Radiata* **11**: 3-20

Vamberger, M., C. Corti, H. Stuckas & U. Fritz. 2011. Is the imperilled spur-thighed tortoise (*Testudo graeca*) native in Sardinia? Implications from population genetics and for conservation. *Amphibia-Reptilia* **32**: 9-25

15. Western Indian Ocean tortoises – dwarfs and giants

About 24 million years ago tortoises reached Madagascar from Africa; different genetic studies give different estimates of the timing, ranging from 17.5 to 24 million years ago. Rapid diversification seems to have followed, giving rise to three genera on Madagascar (*Pyxis, Astrochelys* and *Dipsochelys*) and to tortoises on the western Indian Ocean islands (*Cylindrapsis* and *Dipsochelys*).

The origin of the Madagascar tortoises is not clear; genetically they are close to the group of *Kinixys, Chelonoidis* and *Geochelone*. Fossils from east Africa may represent the ancestral form that crossed the Mozambique Channel to Madagascar. '*Geochelone*' *laetoliensis* from Kenya 4 million years ago was originally described as *Geochelone (Aldabrachelys) laetoliensis* due to a superficial resemblance to the living Indian Ocean giant tortoises, in particular its large size (1 m). It had projecting gulars similar to *Astrochelys* and could have been related to that genus rather than *Aldabrachelys/Dipsochelys*. In contrast to the living Western Indian Ocean tortoises it lacked a nuchal scute and may be closer to *Geochelone*, but it is difficult to interpret the relationships as this species is known only from six very fragmentary specimens. Whether or not *Geochelone laetoliensis* is related to them, the Western Indian Ocean tortoises are probably descended from an African giant tortoise that crossed the Mozambique Channel somewhere around 17 million years ago.

The ancestral large Madagascar tortoise gave rise to the giant *Dipsochelys* and the smaller Malagasy tortoises about 14 million years ago. The back of the skull of these smaller tortoises is largely closed off by the bones closing around the cranial nerves, this is a common process in tortoise evolution although its significance is far from clear, whilst the giants are a very distinctive genus. Most notable of the features of the giants is the front of the skull. The nasal aperture is much higher and more open than any other tortoise and is associated with flaps of skin inside the nose. These act as a valve enabling the animals to draw water through the nose and into the mouth

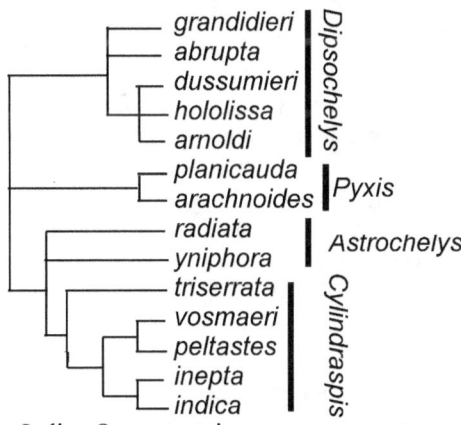

Phylogeny of Western Indian Ocean tortoises

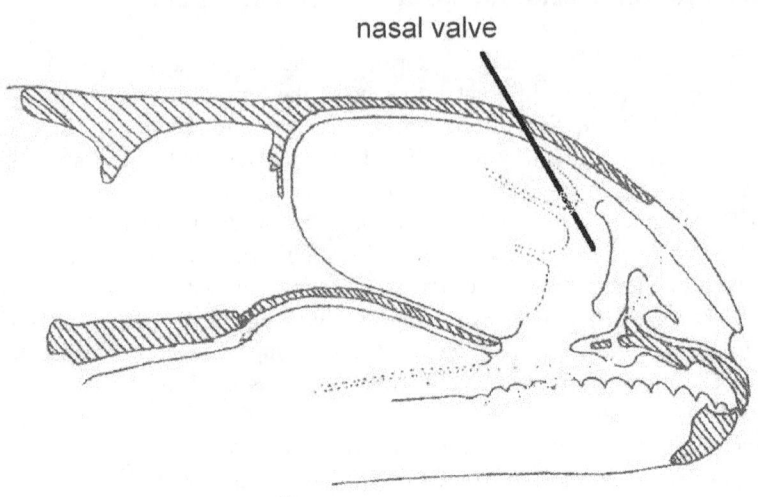

nasal valve

The nasal valve in *Dipsochelys* - tortoise head in cross section, skull bones hatched.

efficiently. The *Dipsochelys* giant tortoises habitually drink in this way, rather than scooping water into their mouths as most tortoises do. This 'nasal drinking' does not require the valves, as the radiated tortoise *Astrochelys radiata* also drinks in this way but has a more normal tortoise nose; however, the *Dipsochelys* system is probably more efficient and may enable high volumes of water to be sucked up relatively quickly. The radiated tortoise is a smaller animal and may therefore easily be able to draw up sufficient water for its size. This feature has been suggested to be an adaptation to the dry conditions on Aldabra atoll, but its presence in the ancestral giants from Madagascar show that it evolved on Madagascar, and was a preexisting feature that enabled the giants to be very successful colonists of Aldabra.

    Dipsochelys is known in Madagascar from two extinct species: *D. abrupta* and *D. grandidieri*. These two species were found in more or less the same parts of western Madagascar and were partially separated ecologically, *D. abrupta* seeming to be associated with open and marsh areas and *D. grandidieri* with forest habitats. The most recent remains have been dated at 750 years ago and extinction may have been due to hunting by humans or to ecological change. Humans arrived in Madagascar 1500 years ago and *D. grandidieri* died out within 450 years of this colonisation, whereas *D. abrupta* survivied a further 300 years. Some *D. grandidieri* shells appear to show man-made holes which supports the idea that humans caused their extinction. These Madagascan giants were the ancestors of the recent giant tortoises from the Seychelles islands. The date of their arrival in the islands is not clear; the living species separated from *D. grandidieri* around 7 million years ago, but their relationship to *D. abrupta* is not known. If they were more closely related to *abrupta* than to *grandidieri* their origin would have been more recent. Two colonisation routes are possible – an easterly route from the south-west of Madagascar, up the coast of Madagascar to the granitic islands or a westerly route up the Mozambique channel, from western Madagascar to Aldabra, with subsequent dispersal north-eastwards to the granitic Seychelles islands. Aldabra has been submerged on a number of occasions over the past two million years (with an unknown earlier history). If tortoises moved from Madagascar to Aldabra they must have colonised the atoll repeatedly. Repeat colonisation via the Mozambique channel from the *D. grandidieri* populations in the south-west is unlikely, but dispersal from *D. abrupta* further to the north may have been more probable. The nearest Madagascar locality with giant tortoise fossils is Amparihingidro where numerous fragments of shells were discovered in a marsh in 1961, lying above a deposit dated at around 2,850 years ago. These have been attributed to *D. abrupta*, a species similar to the living giant tortoises in general shape but slightly larger and considerably more robust. Interestingly, of the fossils from Aldabra, a humerus in the Natural History Museum (collected from Pointe Hodoul) is most similar to Madagascar humeri, although it does not seem to match exactly any of the named forms. This specimen comes from deposits dated at 100,000 years ago. All the other fossil bones from Aldabra are too fragmentary to compare usefully with other material. Tortoises may have drifted from Madagascar repeatedly, most recently 80,000 years ago.

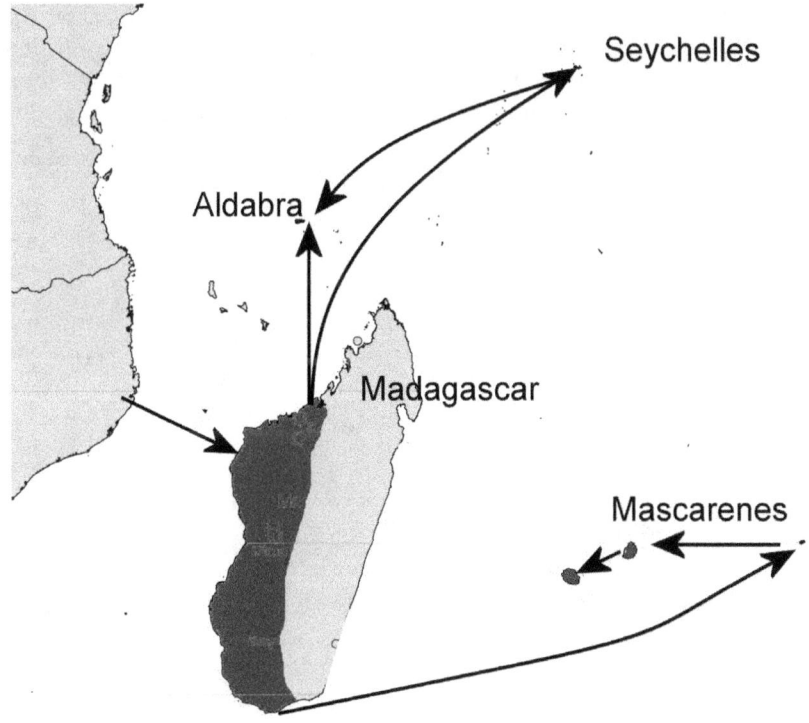

Colonisation routes of Western Indian Ocean species over the past 24 million years.

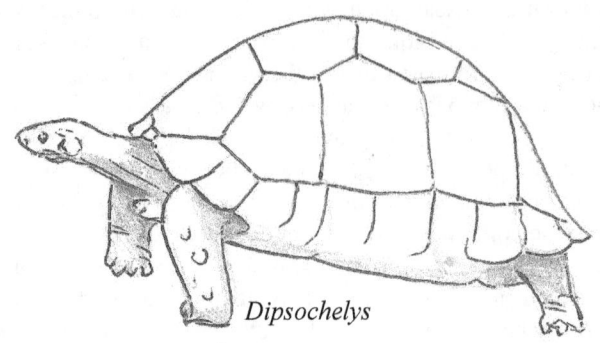

Dipsochelys

The present-day giant tortoises on Aldabra are therefore most likely to have originated either from a population in north-west Madagascar or from the granitic Seychelles islands. This latter possibility would mean that Aldabra provided a stepping-stone to the granitic islands but was then repopulated by granitic island tortoises. This would give a very small gene pool in the original colonisations, an even smaller granitic island gene pool and finally a genetically uniform living Aldabran population. Direct colonisation of the granitic islands from Madagascar could have enabled a moderate level of genetic divergence, with a subset of that gene pool dispersing to Aldabra. The limited genetic data available suggest very low divergence in all the living Seychelles-Aldabra tortoises, supportive of the first, Aldabran stepping-stone hypothesis. Whatever the colonisation route, Seychelles tortoises diverged into two different morphotypes,

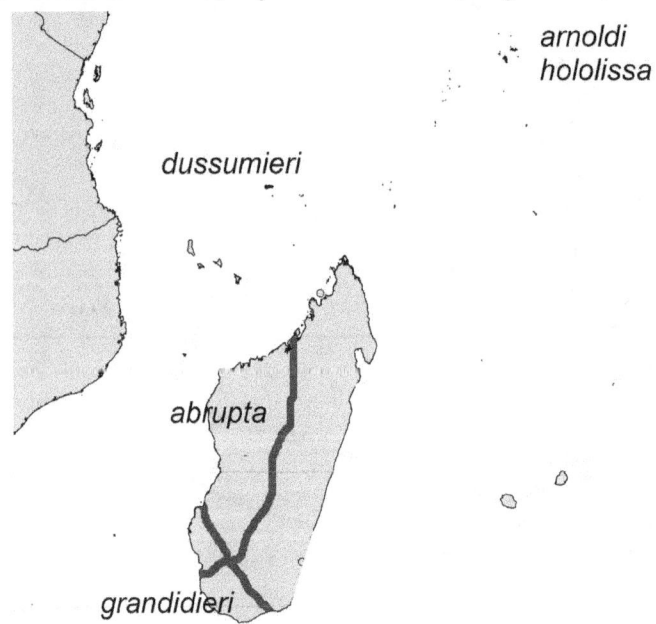

Distribution of *Dipsochelys* populations.

referred to as the Seychelles giant tortoise (*hololissa*) and Arnold's giant tortoise (*arnoldi*), with a third morphotype on Aldabra. The nomenclature of these tortoises is highly contentious, with the Aldabran tortoise being variously referred to as *Aldabrachelys gigantea, Dipsochelys dussumieri* or *D. elephantina*. Which name to apply is currently being decided by the International Commission on Zoological Nomenclature, but while the commission deliberates, a remarkably acrimonious debate continues. All these living forms (and the extinct Daudin's giant tortoise, *daudinii*) are genetically virtually indistinguishable even though their morphological features seem to be the result of distinct development patterns. This may show that they are in the process of evolving into distinct forms but are still at an early stage of the evolutionary process. What little genetic variation there is suggests that this process may have started within the past three million years.

The other Malagasy tortoises seem to have adapted to feeding on the coarse vegetation of the arid south and west of Madagascar, giving rise to animals resembling the medium-sized *Astrochelys*. Some individuals of this type were washed across to the Mascarene islands, there giving rise to the giant *Cylindraspis* tortoises. Others remained on Madagascar, diverging into two genera around 15 million years ago: the dwarf *Pyxis* and the medium-sized *Astrochelys*. This was probably a process of adaptation to microhabitat differences as both forms occur in the same general regions of Madagascar, mainly along the west coast. In both genera there has been a division between a southern scrub species and a more northern, woodland species.

Pyxis contains two species, the flat-tailed tortoise *P. planicauda* and the spider tortoise *P. arachnoides*. The genus was named in reference to the Roman 'pyxis', or jewel box, in reference to the small, ornamented, box-like shell. The dwarf tortoises seem to be essentially paedomorphic (juvenile forms) of the larger species; some subspecies of the spider tortoise show a the juvenile-type hinge of the plastron. This plastral hinge can be moved but is not as mobile as that of the box turtles, rather than a true hinge is a weak line on the plastron retained from the juvenile state. Other juvenile characters retained by the adults are the incomplete bony surround for the stapes, and the incomplete fusion of the upper jaw. It is probable that as the large, possibly even giant, ancestral tortoises on Madagascar started evolving into *Astrochelys* some of the genes regulating development of different parts of the body mutated. This gave rise to a new form that grew slowly

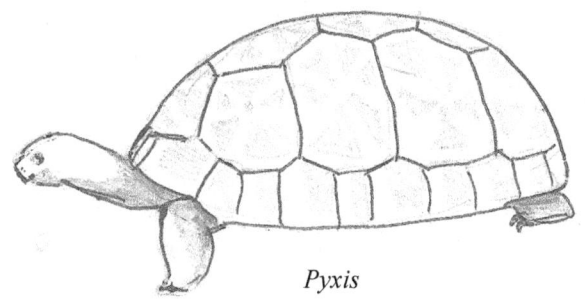

Pyxis

203

but matured very quickly (at less than 10 years old instead of at least 20); adults would then look like juveniles of the ancestors. Following this small genetic change the new diminutive tortoise species would have been selected for adaptations to living in the small spaces available to a small animal. Over millions of years such adaptations may have included becoming even smaller but apparently little changed in morphology from that first early maturing juvenile.

The spider tortoise is only 15 cm long and has a distinctive pattern of yellow star-like markings on black scutes. The plastron may be uniformly yellow or with some dark marks. These tortoises occur in the south-west of Madagascar in semi-arid thornbush. They are only active at cool times, in the early morning and in the rainy season, remaining buried the rest of the time. Three distinct subspecies are recognised: *P. a. brygooi* in a very small area of the northern part of the range; *P. a. arachnoides* in the centre; and *P. a. oblonga* in the south. The three subspecies differ in colour and plastron mobility, from a yellow, immobile plastron in *P. a. brygooi* to a yellow and black plastron with a mobile hinge in *P. a. oblonga*. *P. a. arachnoides* is intermediate in both characters. Genetically *arachnoids* seems to be more distantly related to the other two. *P. a. brygooi* is genetically uniform, whilst there is a small amount of variation in the other subspecies. *P. a. arachnoides* and *P. a. brygooi* hybridise in a narrow band.

The flat-tailed tortoise was placed in a separate genus for many years, but *Acinyxis* is not now considered valid. The two species have been calculated to have diverged around 12 million years ago, but the dating may not be reliable. The flat-tailed tortoise is restricted to a very small area of western Madagascar. It was probably more widespread in the past but is now restricted to the Andranomena forest and nearby areas. The species seems to have a very precise reproductive ecology; captive animals need specific temperature and moisture triggers to stimulate courtship. This suggests that they may be very vulnerable to climate and habitat changes and are restricted to a shrinking ecosystem. It probably originated 12 million years ago when the Madagascan climate was wetter, and forest habitats more widespread along the west coast. The species has been in decline since then as its preferred habitat shrinks at an ever faster rate.

The two *Astrochelys* species share a ridge on the upper jaw at the meeting of the maxillae and premaxillae bones, and the presence of keels on the supraoccipital crest. These are probably connected with feeding specialisations, giving respectively strong ridges on the jaws and strong muscle attachments associated with a relatively powerful bite. Both species inhabit dry habitats; dry forest in the case of the ploughshare tortoise *Astrochelys yniphora* and semi-arid scrub in the case of the radiated tortoise *A. radiata*. In both habitats much of the diet will be made up of coarse vegetation needing these feeding adaptations. Despite its tiny range there is a moderate amount of genetic diversity in the ploughshare, suggesting that it may have been a much more widespread and abundant species in the relatively recent past. A distinct genus, *Angonoka*, was proposed for the ploughshare tortoise in 2006 but this has not been widely accepted and the species is generally included in *Astrochelys*, along with the radiated tortoise. The radiated tortoise is restricted to the semi-arid spiny forest of southern Madagascar. In this area the range has been divided by the Menarandra and Menambovo Rivers, which break the population into three genetically isolated groups. As with the *Pyxis* species,

Astrochelys

the two species may have diverged about 12 million years ago.

Until 200 years ago giant tortoises were present on the Mascarene islands of Mauritius, Rodrigues and Réunion. These Mascarene giants of the genus *Cylindraspis* were descended from an ancestral *Astrochelys*-like tortoise in Madagascar. They share with the radiated tortoise a well developed vertical ridge on the skull (the 'cylindraspid ridge'). Some 10-12 million years ago a tortoise of the radiated lineage was carried by ocean currents to the Mascarene islands. As the earliest species was the Mauritian *C. triserrata* it has been suggested that around 10 million years ago ancestors of this species arrived on the recently formed volcanic island of Mauritius, its descendants then being washed eastwards to Rodrigues, followed by a return westwards to Mauritius and thence to Réunion. This pattern of dispersal backwards and forwards between isolated islands in a vast ocean seems improbable and a simpler scenario can be proposed. 10 million years ago Mauritius was just forming as an active volcano, and would have been uninhabitable. Rodrigues was probably in existence although only recently formed rocks remain on this very eroded island and it cannot be dated properly. It is agreed that

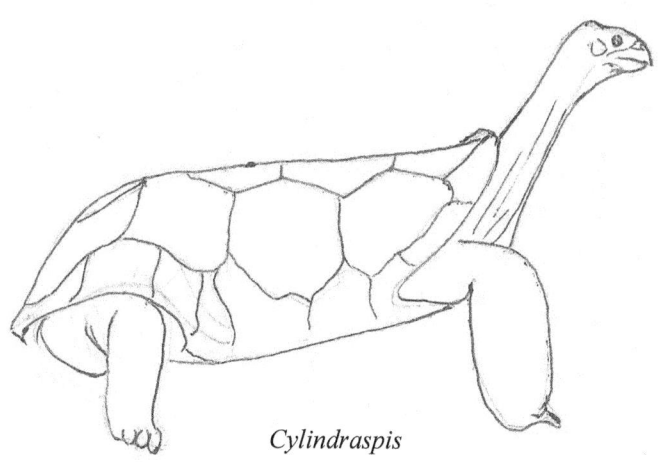

Cylindraspis

a radiated-type tortoise drifted to an island in the Mascarenes 10 million years ago; there it adapted to the island conditions of a highly productive tropical environment and the absence of any competing large herbivores, other than birds such as the dodo. It evolved into the giant *Cylindraspis* tortoises on what is probably the oldest of the Mascarene islands, Rodrigues. Over millions of years the volcanic islands of the Mascarenes have been drifting eastwards, away from the volcanic hotspot. As they move they cease being volcanically active, erode and gradually subside into the sea. The animals and plants on them are carried eastwards as the isloands themselves move, animals falling into the sea are carried by the prevailing ocean currents westwards, back towards the hotspot. 9 million years ago a giant tortoise was carried from Rodrigues to Mauritius, where volcanic activity had subsided and conditions for colonising animals were becoming favourable. There they give rise to *C. triserrata*. 8 million years ago the same process occurred again, giving rise to a second Mauritian species, *C. inepta*. However, there were now two islands in the east and within the past 3 million years tortoises from Mauritius were able to colonise the new island of Réunion, there becoming *C. indica*. The tortoises that remained on Rodrigues diverged into two specialised forms, the giant saddle-backed *C. vosmaeri* and the small domed *C. peltastes*. This scenario is speculative but appears more plausible than the alternative. Ongoing research on the palaeontology of the Mascarenes is may eventually provide the data which could enable us to decide which scenario is correct.

Despite being very little known until recently, *C. indica* is the most important giant tortoise in history. This is not because of what it was, but rather what it was not. The species was named, as *Testudo indica*, in 1783 based on a specimen in the Museum Nationale d'Histoire Naturelle in Paris. This specimen was supposedly from the "Côte de Coromandel" in India, an attribution which prompted the name *indica*. At that time scientific specimens were often labelled with the port they had been shipped from, if they were labelled at all. For the next 72 years many large tortoises were considered to be *Testudo indica*, irrespective of whether they came from Réunion, or other Mascarene

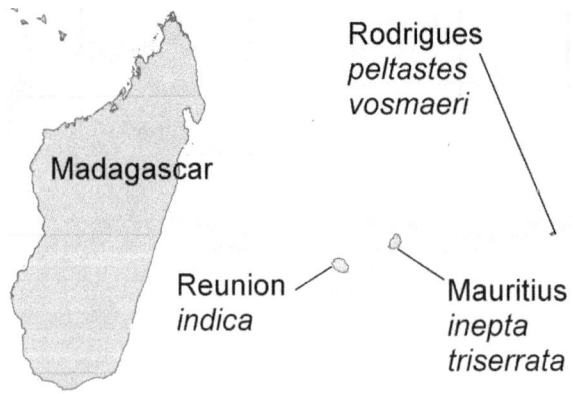

Cylindraspis distribution.

islands, the Seychelles islands, or even the Galapagos. The Réunion tortoises disappeared somewhere between 1775 and 1840. It was not until the 1980s that specimens of known provenance appeared, with the discovery of abundant subfossil remains. These were initially identified as *C. borbonica*, a species named in 1979 by Roger Bour based on drawings made in 1736. DNA analysis subsequently resolved the mystery of the origins of *C. indica* by identifying it (and another mystery species, *C. graii*) as the same species as *C. borbonica*.

Although very few specimens of Mascarene tortoises are useable for genetic studies (18 in total, 2-5 for each species) some genetic variation has been found in some species. This may represent diversification on the different islands, although in *C. indica* this has been speculated to be because of 'incomplete lineage sorting'. This means that several tortoises with different mitochondrial DNA colonised the islands and that insufficient time had elapsed for one form to dominate the others. The idea of multiple tortoises washing from Mauritius to Reunion seems questionable, especially given that none of the mitochondrial forms were found in Mauritian samples. Perhaps the Reunion tortoises were starting to diverge, a possibility which seems to be supported by the presence of four morphological forms on the island. These may correspond to males and females of two different types. The different samples for *C. inepta* are found on different islands, with one specimen from Mauritius being different from the two specimens from Isle aux Aigrettes. These levels of genetic variation could suggest divergence over a million years. Whether these were just genetic forms or had started to vary in appearance is unknown because of the very fragmentary nature of the bones recovered from caves and marshes.

Morphologically the species diverged in that the genus exaggerated the feeding specialisations already present in their radiated tortoise ancestors. The cylindraspid ridge for jaw muscle attachment expanded and the jaws developed tooth-like ridges and processes. In addition a third ridge developed on the lower jaw, all contributing to a strong bite and grinding jaw action, suggestive of adaptation to feeding on coarse vegetation. Whilst much of the Mascarene islands was covered in lush tropical forests before the arrival of humans, one of the most extensive habitats in the lowlands was palm savannah, a habitat type now almost completely lost. This may have been the most suitable habitat for tortoises, meaning that a substantial part of the diet may have been coarse grasses and palm leaves. Further adaptations were reduction in shell thickness, which has been suggested to have been made possible by the absence of predators of adult tortoises on the islands. The thin shell would have been structurally weak which in turn may have necessitated the fusion of shell bones. *C. triserrata* may have been the most extreme dietary specialist, having evolved an additional third ridge on the upper jaw. It also reduced the inguinal scute, and developed paired, diverging gulars unlike most other Western Indian Ocean tortoises which have paired but not diverging gulars and other Mascarene tortoises which started out with paired gulars but fused them early in development. The others all share possible tendencies towards a saddle-backed browsing form in that the plastron is reduced, the openings for the limbs, and in particular the neck, widened, modification of the fore-limbs in having one of the processes on the head of the femur lowered, and a flattened ridge on the skull associated with the jaw muscles. The two Rodrigues species have a change in the size of the arteries in the head, the foramina in the skull for passage of the arteries show that the palatine circulation was more developed than the temporal. What this meant to the animals is not known. *C. vosmaeri* was saddle-backed, *C. peltatses* small and domed, with a short supraoccipital crest suggestive of retention of some juvenile characters. The Mauritian-Réunion species pair *C. indica* and *C. inepta* had fused skulls and broad snouts. *C. inepta* was a domed species whereas *C. indica* was saddle backed, and seems to have been adapted to less coarse vegetation with reduced processes at the back of the skull for muscle attachment and loss of the tooth-like projections on the jaws.

Whilst the *Dipsochelys* giants of Seychelles are the most abundant giant tortoises and the *Cylindraspis* giants of the Mascarenes were probably the strangest and most interesting of the giant tortoises, it is the giants of the Galapagos islands that have become most famous. Their story is one of ocean crossing from its very beginning.

References

Austin, J. & E.N. Arnold. 2001. Ancient mitochondrial DNA and morphology elucidate an extinct island radiation of Indian Ocean giant tortoises (*Cylindraspis*). *Proc. R. Soc. Biol. Sci. Ser. B* **268**: 2515-2523

Bour, R., 1994. L'etude des animaux doublement disparus; les tortues géeantes subfossiles de Madagascar. *Mern. Trav. Inst. Montpellier* **19**: 1-253

de Broin, F. 1990. Tortues. In : Allibert et al. Le site de Dembeni, *Etudes Ocean Indien* (11) **1**: 63-172

Caccone, A., Amato, G., Gratry, O. C., Behler, J. & Powell, J. R. 1999. A molecular phylogeny of four endangered Madagascar tortoises based on mtDNA sequences. *Mol. Phyl. Evol.* **12**: 1-9

Chiari, Y., M. Thomas, M. Pedrono & D.R. Vietes. 2005. Preliminary data on genetic differentiation within the Madagascar spider tortoise, *Pyxis arachnoides* (Bell, 1827). *Salamandra* **41**: 35-43

Gerlach, J. 2004. *Giant Tortoises of the Indian Ocean*. Chimiara publishers, Frankfurt.

Harrison, T. 2011. Tortoises (Chelonii, Testudinidae). In: T. Harrison (ed.), *Paleontology and Geology of Laetoli: Human Evolution in Context. Volume 2: Fossil Hominins and the Associated Fauna*, Vertebrate Paleobiology and Paleoanthropology, DOI 10.1007/978-90-481-9962-4_17

Le, M., C.J. Raxworthy, W.P. McCord & L. Mertz. 2006. A molecular phylogeny of tortoises (Testudines: Testudinidae) based on mitochondrial and nuclear genes. *Mol. Phyl. Evol.* **40**: 517-531

Palkovacs. E.P., J. Gerlach & A. Caccone. 2002. The evolutionary origin of Indian Ocean tortoises (*Dipsochelys*). *Mol. Phylogenet. Evol.* **24**(2): 216-27

Palkovacs, E.P., M. Maschner, C. Ciofi, J. Gerlach & A. Caccone. 2003. Are the native giant tortoises from the Seychelles really extinct? *Mol. Ecol.* **12**: 1403-1414

Rioux Paquette, S., S.M. Behncke, S.H. O'Brien, R.A. Brenneman, E.E. Louis Jr. & F.-J. Lapointe. 2006. Riverbeds demarcate distinct conservation units of the radiated tortoise (*Geochelone radiata*) in southern Madagascar. *Conserv. Genet.* **8**: 797-807

Walker, R.J. 2010. The decline of the Critically Endangered northern Madagascar spider tortoise (*Pyxis arachnoids brygooi*). *Herpetologica* **66**(4): 411–417

16. South American tortoises – colours and more giants

30 million years ago a large tortoise form was present in Africa and southern Asia. This has given rise to the South American *Chelonoidis*, the Indian *Geochelone* and the African *G. (Centrochelys) sulcata*. At this time South America was completely isolated, having lost its original contact with Africa 180 million years ago and still being hundreds of kilometres away from the North and Central American land mass. This means that the appearance of tortoise fossils in South America 24 million years ago and molecular clock calculations of their origin at exactly the same time must have resulted from colonisation through dispersal over the ocean.

Dispersal from Africa to South America probably occurred by tortoises floating, by themselves, or on rafts of vegetation washing out from rivers in the Congo and west Africa, and crossing the Atlantic on the strong equatorial currents to Brazil. Although this seems unlikely it is the only plausible explanation for the origin of South American tortoises and for the continent's monkeys (separating from their African ancestors around 35 million years ago), large skinks (two colonisations in the last 9 million years) and rodents. Tortoises are well adapted to floating across oceans with their buoyant structure and slow metabolism. Colonisation of South America from Africa may have been facilitated by the presence of islands between the continents (now submerged); these islands on the Ceara and Sierra Leone Rises would have been exposed 30 million years ago when sea levels were considerably lower.

However the ancestral tortoise crossed the Atlantic, by 24 million years ago it had reached South America. The earliest South American fossils have been placed in '*Geochelone*' but are of uncertain identity; they may be either *Geochelone* or *Chelonoidis*. The South American genus *Chelonoidis* appears to have been distinct from *Geochelone* by 15 million years ago.

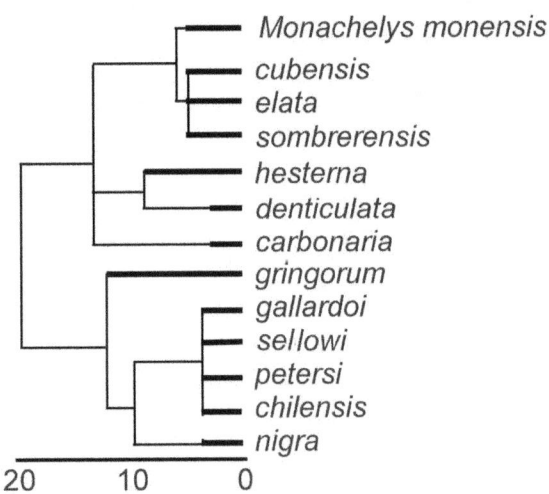

Phylogeny of *Chelonoidis*. Horizontal scale in millions of years ago

There appears to have been an early diversification of tortoises in South America, although only a small number of species resulted. All living species are in the genus *Chelonoidis*, with three or four mainland species which are all descended from a common ancestor around 14 million years ago. The oldest definite members of the genus are *C. hesterna* from Colombia and *C. gringorum* from Argentina, from 13 and 15 million years ago respectively. *C. hesterna* may be ancestral to the living yellow-footed tortoise *C. denticulata* and the red-footed tortoise *C. carbonaria*. *C. gringorum* may be ancestral to the Chaco tortoise *C. chilensis*.

Of the living mainland species the red-footed tortoise is the most widespread, from Panama south to Argentina. This species has been introduced to many Caribbean islands and to Nicaragua. It contains several distinct mitochondrial groups. The southern genetic group split from the main population around 4 million years ago, followed by the northern group 3 million years ago and a final split between the north-western, north-eastern and eastern groups 2 million years ago. This sort of splitting is often attributed to habitat changes due to climates altering in the Ice Ages, but in this case they occurred before the start of Ice Age conditions. The red-footed tortoise's preference for open grassy habitats gives it a fragmented distribution in the largely forested northern half of South America; during the Ice Ages, dry, cooler conditions would have been expected to result in expansion of these grasslands. However, if this did occur it was insufficient to allow these grasslands to join up into a continuous area and the red-footed tortoise populations have remained isolated from one another.

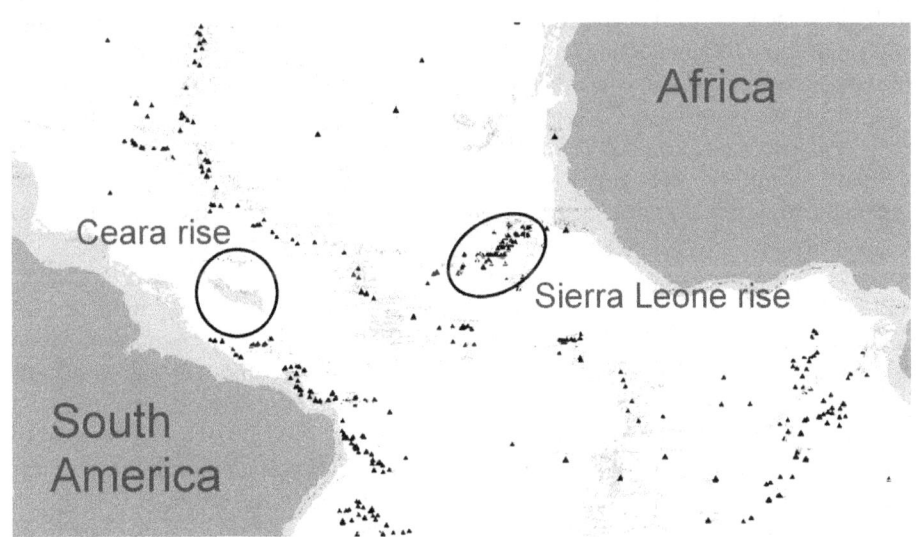

Ceara and Sierra Leone rises in the Atlantic Ocean. Shallow water areas that may have been exposed as dry land 30 million years ago shown in grey. Sea-mounts are shown in black; some of these may have formed islands in the past.

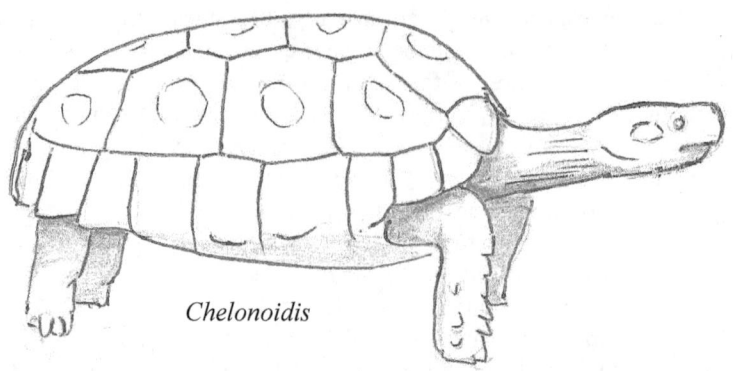

Chelonoidis

The red-footed and the yellow-footed tortoise share similar general distributions in northern South America, but whereas the red-footed is a savannah and dry forest species, the yellow-footed is found in humid forests. These forests have changed in area over the past two million years. There has been some change in distribution with a 120,000 year old fossil of this species found 800 km south of the present range. However, corridors seem always to have existed between the main forest patches and the yellow-footed tortoise populations have not been fragmented; as a result there are no clear genetic groups within this species. These two large South American tortoises were essentially separated by their specialisation to different habitats. They differ in some aspects of shape and size, but mainly in colouration; the distinctive red and yellow colours being found on the feet (as the names indicate) and head. The bright head markings are used in courtship displays, which may reinforce the separation.

Peters' tortoise *C. petersi* ranges from Bolivia to northern Argentina. This may be synonymous with the Chaco tortoise which is known from fossils from within the past 2.5 million years to the present day in Argentina. Other fossil species from within the past 5 million years from the area may also be related to the Chaco tortoise: *C. gallardoi* from Argentina and *C. sellowi* from Uruguay.

Tortoises rafted or floated to the West Indies around 2 million years ago, and probably on several occasions since. The early fossils are mostly represented by fragmentary remains and the species are poorly known. They have been found on Mona Island off Puerto Rico (*Monachelys monensis*) and the nearby Turks and Caicos islands, Cuba (the giant *Chelonoidis cubensis* and *C. elata*), Sombrero Island (*C. sombrerensis*) Navassa, Curacao, Barbados and Antigua. As is common in island tortoises, these were mostly large animals; the possibly 2 million year old *Monachelys* was about 50 cm long and *C. cubensis* was a giant at over 1 m. There are more recent (15,000 years old) fossils from Hispaniola; these seem to have been adapted to very dry conditions on the island, being restricted to dry coastal areas as conditions became wetter inland, and disappearing by 13,000 years ago. Whether or not this habitat change caused the extinction of all the other species is not known.

In the north of the Caribbean (Bahamas) the extinct tortoises are attributed to the North American genus *Hesperotestudo*, but those from all the other, more southerly,

212

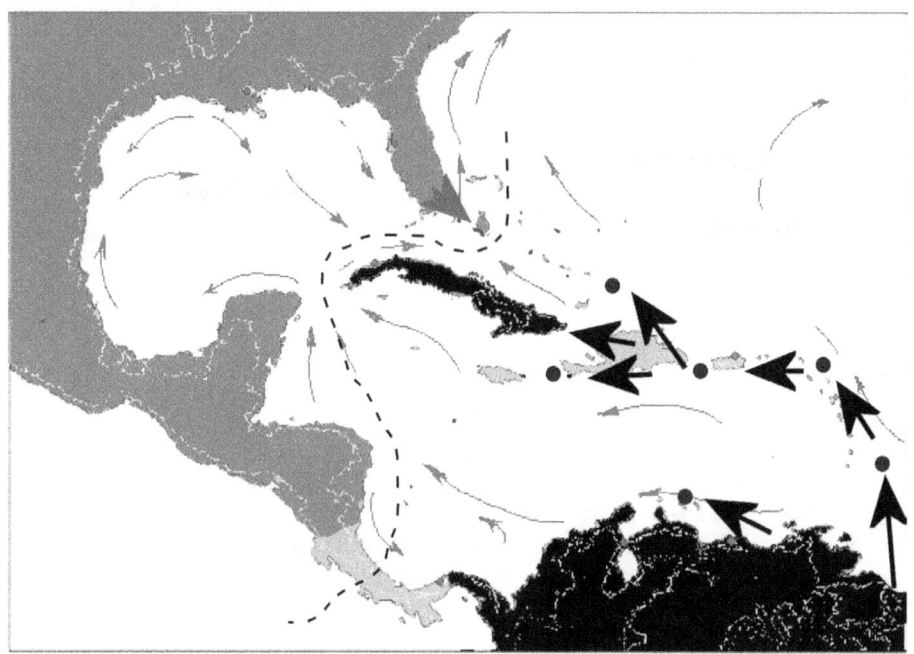

Caribbean tortoise distribution, *Chelonoidis* and *Monachelys* south of the dashed line (in black), *Gopherus* and *Hesperotestudo* lineage north of the dashed line (in dark grey).

islands appear to be closer to *Chelonoidis*. Thus the Caribbean islands supported two very different lineages of tortoise in the past two million years ago but nowhere did these lineages meet. Five million years ago the Panama Isthmus formed, allowing North and South American faunas to interact. *Chelonoidis* moved north into Panama but the surviving North American tortoises were all associated with arid habitats and failed to move south. Today there remains a gap of at least 3,000 km between these tortoises.

On the west coast of South America rafting gave rise to the colonisation of the Galapagos islands. These famous giant tortoises form the *Chelonoidis nigra* complex of species or subspecies, which have also been placed in a distinct (but now rejected) genus *Elephantopus*. The closest living relative to the Galapagos tortoises is the Chaco tortoise. The ancestor of the Galapagos tortoises probably drifted to the Galapagos on the strong ocean current up the west coast of South America. It may have drifted to the Galapagos and evolved large size on the islands, or may have given rise to a larger form which would have been more likely to survive the extended drifting to the islands. Molecular data suggest that the Chaco and Galapagos tortoises diverged around 9 million years ago. Although the oldest of the existing islands is no more than 5 million years old, the existing Galapagos islands lie on top of a volcanic hotspot which has been present for some 10 million years. A chain of sea mounts represent these of older islands, now lying below the sea to the east of the existing islands. Islands have probably been present at the Galapagos volcanic hotspot for at least 10 million years, giving ample time for South American tortoises to drift to the islands.

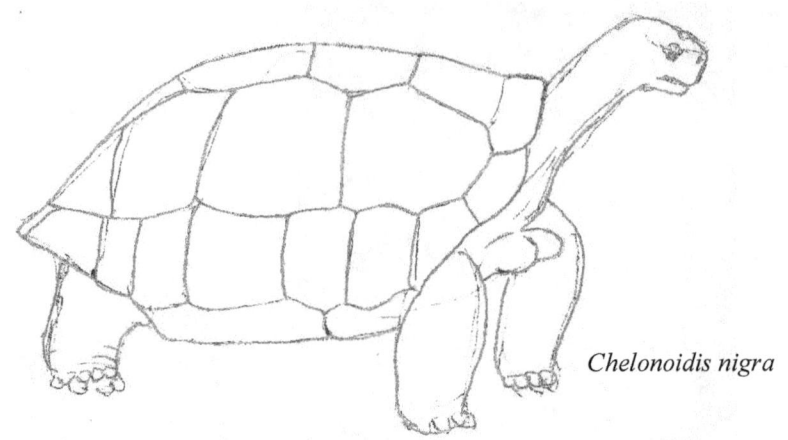

Chelonoidis nigra

The original diversity of tortoises on the Galapagos islands is somewhat confused as a result of the early extinction of some populations, confused historical records and the movement of tortoises between localities. Currently 11 named forms are recognised and a further un-named form from Santa Cruz is recognised as distinct:

Chelonoidis abingdonii Abingdon or Pinta Island giant tortoise
C. becki Volcan Wolf giant tortoise (Isabella island)
C. chathamensis Chatham or San Cristóbal Island giant tortoise
C. darwini James or Santiago Island giant tortoise
C. duncanensis Duncan or Pinzón Island giant tortoise
C. hoodensis Hood or Española Island giant tortoise
C. nigra Charles or Floreana Island giant tortoise
C. phantastica Narborough or Fernandina island giant tortoise
C. porteri Indefatigable or Santa Cruz Island giant tortoise
C. vandenburgi Volcan Alcedo giant tortoise (Isabella island)
C. vicina Isabella Island giant tortoise (Isabella island)
C. wallacei Rabida Island giant tortoise

The oldest of the existing Galapagos islands are in the east, with the younger, volcanically active islands in the west. The oldest of the islands is Española at 3.3 million years old. This island could be the source of all the living Galapagos tortoises which seem to share a common ancestor no more than 3.2 million years ago. The evolutionary scenario for the Galapagos makes more sense though if their ancestor is assumed to have lived on a now submerged island to the east of Española. Tortoises drifting from that island on the westward currents that pass though the islands would have been carried to Española, giving rise to the Hood Island giant tortoise *C. hoodensis.*

As Española is one of the most accessible and hospitable of the Galapagos islands (its age meaning that it has no active volcanoes and the steep volcanic mountains have eroded to a relatively flat topography), the island was drastically affected by the

arrival of humans. By the mid-1960s
the few remaining tortoises had stopped
breeding and the last wild animals
(two males and 12 females) were taken
into captivity at the Charles Darwin
Research Station. The first hatchlings
were produced in 1971 and released
on Española four years later. In 1977

Chelonoidis hoodensis

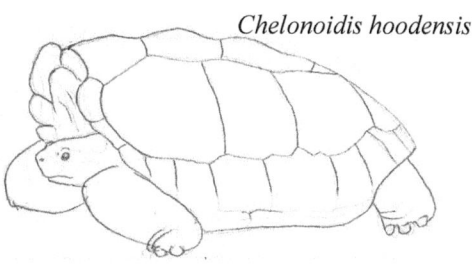

another male (from San Diego Zoo) was added to the breeding group, resulting in an
increase in breeding activity. By 2002 1200 juveniles had been returned to Española. By
that time the island was a much safer environment for tortoises, with legal protection
and the eradication of the goats that had destroyed much of the vegetation; the
environment is still somewhat degraded in that the *Opuntia* cactus on the island has not
fully recovered from the decades of goat browsing. In 1988 the first nests produced by
repatriated tortoises were found and the first Española hatched juveniles were located in
1994. The 1200 Española tortoises that have been reintroduced are descended from the
three males and 12 captive females, with the majority descended from one of the males.
This species is still considered to be critically endangered.

Around 1.7 million years ago the Española form seems to have given rise to
tortoise populations on Santa Cruz. Most Santa Cruz tortoises are found in the north and
south-west of the island where two distinct forms occur. A third form in the east, the
Cerro Fatal population seems to have a separate origin (from San Cristobal).

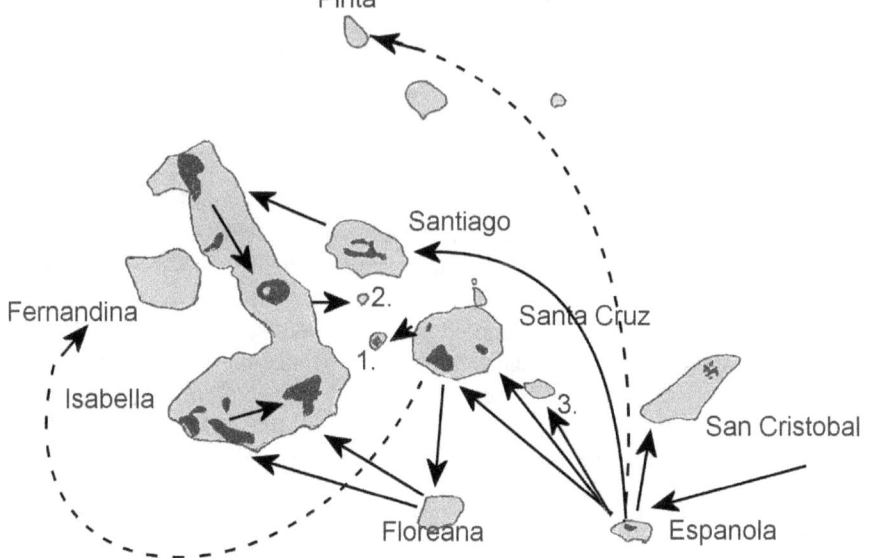

1. Pinzon   2. Rabida   3. Santa Fe

Colonisation patterns of Galapagos tortoise popualtions. Living populations shown in
dark grey.

215

The descendants of the colonists from Española diverged into two distinct populations: the northern Cerro Montura population and the south-western La Caseta (or La Reserva) population. The tortoises from Cerro Fatal and La Caseta have both been ascribed to the Indefatigable Island giant tortoise *C. porteri*, whilst the Cerro Montura form remains un-named. The Cerro Fatal population is less variable than the La Caseta one; this reflects the fact that it was reduced to about 100 animals by 1974, whereas

Chelonoidis porteri

there are still some 2-3,000 tortoises at La Caseta. Some individuals appear to be hybrids, as indicated by genetic data; these rare individuals (3% of studied animals) are probably the result of occasional migration between the areas of the island. The Cerro Montura population crossed to Pinzón island, giving rise to the Duncan Island giant tortoise *C. duncanensis* around 1.2 million years ago. This species is reduced to only 150-200 animals.

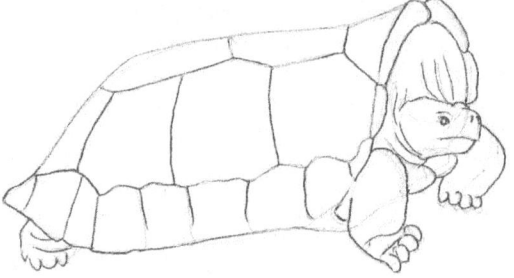

Chelonoidis duncanensis

Española tortoises seem also to have drifted to Santiago 1.4 million years ago. The James Island giant tortoise *C. darwini* form that evolved there is restricted to just 500-700 survivors. This species subsequently spread to northern Isabella within the past 280,000 years; the Volcan Wolf giant tortoise *C. becki* from there is genetically indistinguishable from the Santiago tortoises and morphologically similar. The Volcan Wolf species is classed as vulnerable, with a relatively healthy population of 1-2,000 animals.

850,000 years ago tortoises from La Caseta drifted south to Floreana. This is the Charles Island giant tortoise *C. nigra* which was hunted to extinction in about 1850. From Floreana they drifted to the south of the large island of Isabella 650,000 years ago. In the south of the island they are poorly differentiated and can all be ascribed to the Isabella Island giant tortoise *C. vicina*. Several taxa have been described from this area; the first to evolve form seems to

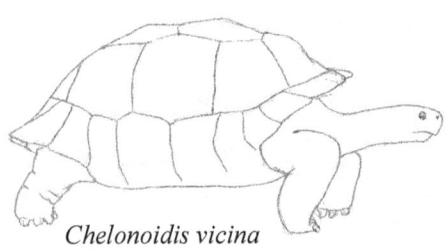

Chelonoidis vicina

216

microphyes, followed by a grouping of *vandenburghi* and a mixture of tortoises described as *guntheri* and *vicina/guntheri*. This population is reduced to 4-6,000 animals and the Isabella Island giant tortoise *C. vicina* is considered vulnerable.

820,000 years ago Española tortoises colonised Santa Fe. This population is extinct; the last record of tortoises on the island was from 1890. This species is known only from some bones and was never named.

San Cristobal island in the east is 3.2 million years old and was probably colonised from Española 650,000 years ago. The island is now occupied by the Chatham Island giant tortoise *C. chathamensis.* This species is vulnerable with only 500-700 animals. 430,000 years ago tortoises drifted from San Cristobal to the east coast of Santa Cruz, resulting in a localised population in the Cerro Fatal area.

Since these different island populations formed there has been movement of some tortoises between the populations on different parts of Isabella, giving some mixing of genes between northern and southern tortoises, mostly due to male tortoises from the north spreading northern genes southwards. However, there has been little mixing for the past 88,000 years. The northern Alcedo tortoise *C. vandenburghi* may be distinctive. This is the largest population of tortoises in the Galapagos (3-5,000 tortoises) but has lower genetic diversity than other Isabella populations. There was a major eruption of Alcedo about 100,000 years ago, which must have destroyed most of the tortoise habitat, and reduced the tortoise population. This is reflected in the tortoise genetics with 91% of Alcedo tortoises being descended from one female. *C. guntheri* has been described from Sierra Negra but seems to be synonymous with *C. vicina*, as does *C. microphytes* from Volcan Darwin. Volcan

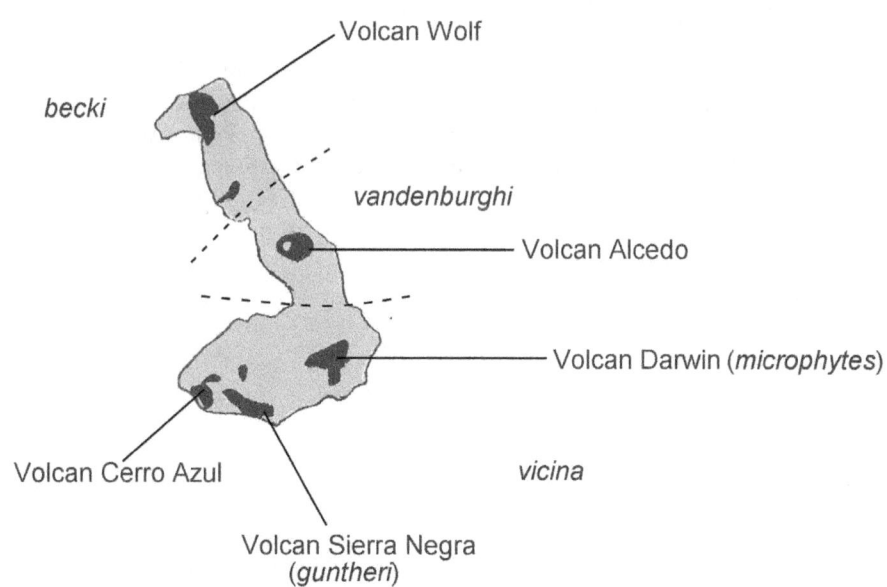

Isabella island, tortoise populations shaded.

Darwin mitochondrial genes are descended from those from Alcedo, with some unique forms. Cerro Azul and Sierra Negra have much overlap. There is also some overlap between these and Alcedo. Mitochondrial gene distributions indicate that there has been dispersal from La Cazuela in the east to Cerro Azul and reverse dispersal from Cerro Azul eastwards to Cinco Cerros, Cabo Rosa and Roca Union. Cabo Rosa tortoises have been described as *C. guntheri*, Cinco Cerros has individuals of both *guntheri* and *vicina*. These rather variable looking tortoises are probably best regarded as *C. vicina*. All of this suggests significant movement between the different south Isabella populations in the past. South Isabella tortoises may also have migrated back to Santa Cruz, resulting in a close nuclear DNA similarity.

The most recent island to be colonised was Rabida, which was colonised from Isabella 30,000 years ago. This population (*C. wallacei*) is now extinct. There are no early records from Rabida, and it has been suggested that the single specimen and isolated bones from there were introduced from another island. DNA from there groups with southern Isabella tortoises which is plausible for both natural colonisation and human transport. Another extinct population is that of Fernandina, described as the saddlebacked Narborough island giant tortoise *C. phantastica*. It became extinct in about 1960 and is known only from the single specimen collected in 1906. DNA from this specimen is very close to that of a south or western Santa Cruz tortoise, although not exactly the same as any of the living Santa Cruz groups, making the origin of this specimen questionable. The single specimen is the only evidence of a tortoise ever having lived on the island; there are no historical records and no bones, and no obvious reason for tortoises to have died out on this, the most pristine of the Galapagos islands. Were the Fernandina tortoises colonists from Santa Cruz, or was the single specimen transported there by humans? With only the single specimen these questions will remain unanswerable.

Another puzzle is that of the Pinta island tortoise. Some 300,000 years ago the Española tortoises seem to have given rise to the Pinta population. This is the most famous of all giant tortoises as it is represented by the single surviving Abingdon Island giant tortoise *C. abingdonii*, Lonesome George. The five tortoises collected on Pinta in 1906 seemed to be the last from that island until one male was seen in 1971. In 1972 the same tortoise was found again and captured. Lonesome George is now in the captive breeding centre of the Charles Darwin Research Station. As he seems to be the last of his species attempts have been made to

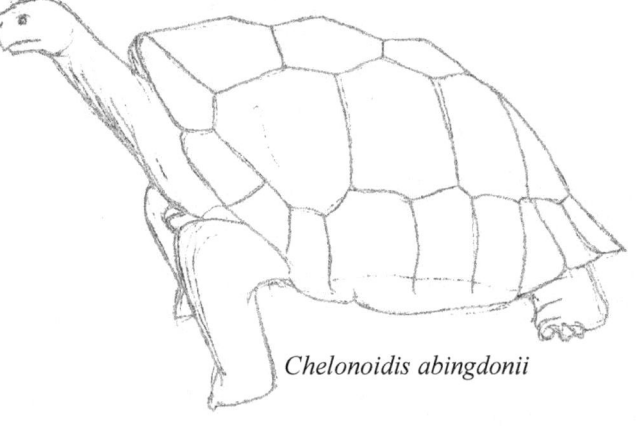

Chelonoidis abingdonii

encourage him to hybridise with females of other species, but this has been a failure so far. Pinta is one million years old but the molecular data suggest a late colonisation from Española. This is surprising as Española and Pinta are widely separated islands. It has been suggested that ocean currents make colonisation of Pinta from Española plausible; this would be expected to produce a close relationship between the populations of these islands but this is not the case. Lava lizards which must have moved by rafting, and so should show similar patterns to floating giant tortoises, seem to have a Pinta population that is descended from Isabella, as would be expected on the grounds of proximity.

The patterns described above were established by mitochondrial DNA and largely supported by nuclear DNA, however they are contradicted by microsatellite DNA studies. These suggest a Santa Cruz origin for all the tortoises, giving rise to San Cristobal and to Isabella. San Cristobal then seems to be the origin of Española and ultimately Pinta while Isabella gives rise to Santiago. Much of this reverses the pattern of the other studies. The interpretations of microsatellites are based on samples of known origin which define genotypes; 'likelihood' methods are then used to interpret samples of unknown origin. Although microsatellites are useful in suggesting the origin of some tortoises the pattern of movement indicated by them in this case seems to be contradictory and perhaps unreliable.

A frequent point of interest in the Galapagos tortoises is the origin of the different tortoise shell types – the 'saddle-backs' and the domed tortoises. Many of the tortoises have flattened shells, often turning up at the front. These saddle-backed shells enable the tortoises to stretch their necks upwards more than most tortoises. This has been assumed to be an adaptation to facilitating browsing on high vegetation for tortoises on islands with very poor quality vegetation; those in areas with lush vegetation could obtain all the nutrition they need by grazing. The high shell opening also enables males to stretch their necks upwards in intimidatory displays; males seem to sort out their dominance hierarchies by seeing who is tallest, competing to stretch their necks higher than their rivals. Whether feeding or sexual competition is the main driving force is unknown, perhaps it depends on whether it is a wet year with plenty of food and or a harsher one. The mitochondrial studies suggest that the earliest Galapagos tortoises were saddlebacked and that domed tortoises have evolved three times (La Caseta, Cerro Fatal and in some south Isabella) with one return to saddlebacks (Floreana). The nuclear DNA (which will contain whatever genes control shell shape) suggests a domed origin and saddlebacks evolving three times (Cerro Montura, some Isabella and Española) and no reversals. Both sets of data give partial change in shell shape in the intermediate population of San Cristóbal. The nuclear DNA seems to make more sense here as it involves fewer changes and suggests a more straightforward domed origin. The shape of the domed tortoises would seem to be better adapted to floating and it seems likely that saddlebacks evolved in isolation on many of the islands rather than dispersing between them.

In addition to the movement of tortoises over the past one or two million years there seems to have been significant levels of more recent movement as indicated by the 'wrong' genotypes turning up in some populations. Of the tortoises sampled on Española, one may be descended from a male from Pinzon. From Floreana, one tortoise may originate from Pinzon, one from Santa Cruz and two from Isabella. One of the sampled animals on Volcan Wolf on Isabella appears to be a first generation hybrid between a Pinta male and a female which may itself have been a hybrid between a Volcan Wolf male and an Española female. Seven possible second generation hybrids between an Española male and Volcan Wolf females have also been found in the same area. In addition two seemed to have Floreana female ancestry. This may be due to human movement of tortoises or to natural dispersal.

The Galapagos tortoises have become symbolic of the plight of tortoises in the modern world: the extinction of three wild populations within 200 years of humans settling the islands highlights how vulnerable these animals are.

References

Auffenberg, W. 1967. Notes on West Indian Tortoises. *Herpetologica* **23**(l): 34-44

Beheregaray, L.B., C. Ciofi, D. Geist, J.P. Gibbs, A. Caccone & J.R. Powell. 2003. Genes Record a Prehistoric Volcano Eruption in the Galapagos. *Science* **302**: 75

Beheregaray, L.B., J.P. Gibbs, N. Havill, T.H. Fritts, J.R. Powell & A. Caccone. 2004. Giant tortoises are not so slow: Rapid diversification and biogeographic consensus in the Galápagos. *PNAS* **101**: 6514-6519

Burns, C.E., C. Ciofi, L.B. Beheregaray, T.H. Fritts, J.P. Gibbs, C. Marquez, M.C. Milinkovitch, J.R. Powell & A. Caccone. 2003. The origin of captive Galápagos tortoises based on DNA analysis: implications for the management of natural populations. *Animal Conservation* **6**: 329-337

Caccone, A., G. Gentile, J.P. Gibbs, T.H. Fritts, H.L. Snell, J. Betts & J.R. Powell. 2002. Phylogeography and history of giant Galapagos tortoises. *Evolution* **56**: 2052-2066.

Caccone, A., G. Gentile, C.E. Burns, E. Sezzi, W. Bergman, M. Ruelle, K. Saltonstall, & J.R. Powell. 2004. Extreme difference in rate of mitochondrial and nuclear DNA evolution in a large ectotherm, Galápagos tortoises. *Mol. Phyl. Evol.* **31**: 794-798

Caccone A., J.P. Gibbs, V. Ketmaier, El. Suatoni & J.R. Powell. 1999. Origin and evolutionary relationships of giant Gálapagos tortoises. *PNAS USA* **96**: 13223-13228

Carranza, S. & E.N. Arnold. 2003. Investigating the origin of transoceanic distributions: mtDNA shows *Mabuya* lizards (Reptilia, Scincidae) crossed the Atlantic twice. *Syst. Biodiv.* **1**: 275-282

Chiari, Y., C. Hyseni, T.H. Fritts, S. Glaberman & C. Marquez. 2009. Morphometrics Parallel Genetics in a Newly Discovered and Endangered Taxon of Galapagos Tortoise. *PLoS ONE* **4**(7): e6272.

Ciofi, C., M.C. Milinkovitch, J.P. Gibbs, A. Caccone & J.R. Powell. 2002. Microsatellite analysis of genetic divergence among populations of giant Galápagos tortoises. *Mol. Ecol.* **11**, 2265-2283

Ciofi, C., K.A. Wilson, L.B. Beheregeray, C. Marquez, J.P. Gibbs, W. Tapia, H.L. Snell, A. Caccone & J.R. Powell. 2004. Phylogeographic history and gene flow among giant Galápagos tortoises on southern Isabella island. *Genetics* **172**: 1727-1744

Franz, R. & & C. Woods. 1983. A fossil tortoise from Hispaniola. *J. Herpetol.* **17**: 79-81

Manzano, A.S., J.I. Noriega & W.G. Joyce. 2009. The tropical tortoise *Chelonoidis denticulata* (Testudine: Testudinidae) from the Late Pleistocene of Argentina and its paleoclimatological implications. *J. Paleont.* **83**: 975-980

Milinkovitch, M.C., D. Monteyne, J.P. Gibbs, T.H. Fritts, W. Tapia, H.L. Snell, R. Tiedemann, A. Caccone & J.R. Powell. 2004. Genetic analysis of a successful repatriation programme: giant Galapagos tortoises. *Proc. R. Soc. Lond.* B **271**: 341-345

Milinkovitch, M.C., D. Monteyne, M. Russello, J.P. Gibbs, H.L. Snell, W. Tapia, C. Marquez, A. Caccone & J. R. Powelll. 2007. Giant Galápagos Tortoises: Molecular Genetic Analysis Reveals Contamination in a Repatriation Program of an Endangered Taxon. *BMC Ecology* **7**: 2

Poulakakis, N., S. Glaberman, M. Russello, L.B. Beheregaray, C. Ciofi, J.R. Powell, & A. Caccone. 2008. Historical DNA analysis reveals living descendants of an extinct species of Galápagos tortoise. *PNAS* **105**: 15464-15469

Pritchard, P.C.H. 1996 The Galapagos tortoises: nomenclatural and survival status. *Chelonian Res. Monogr.* **1**: 1-85.

Russello, M., L.B. Beheregaray, J.P. Gibbs, T. Fritts, N. Havill, J.R. Powell & A. Caccone. 2007. Lonesome George is not alone among Galápagos Tortoises. *Current Biology* **17**: 317-318

Russello, M., S. Glaberman, C. Marquez, J.R. Powell & A. Caccone. 2005. A novel taxon of Giant tortoises in conservation peril. *Biology Letters* **1**(3): 287-290

Russello, M.A., N. Poulakakis, J.P. Gibbs, W. Tapia, E. Benavides, J.R. Powell & A. Caccone. 2010. DNA from the past informs ex situ conservation for the future: an "extinct" species of Galápagos tortoise identified in captivity. *PLoS One* **5**(1): e8683

Vargas-Ramírez, M., J. Maran & U. Fritz. 2010. Red- and yellow-footed tortoises, *Chelonoidis carbonaria* and *C. denticulata* (Reptilia: Testudines: Testudinidae), in South American savannahs and forests: do their phylogeographies reflect distinct habitats? *Org Divers Evol* DOI 10.1007/s13127-010-0016-0

Williams, E.E. 1952. A new fossil tortoise from. Mona Island, West Indies, and a tentative arrangement of the tortoises of the world. *Bull. Amer. Mus. Nat. Hist.* **99**: 541-560

17. A look to the future

The evolution of turtles, terrapins and tortoises described here has occurred over 210 million years. This makes the turtles an ancient group of vertebrates, the first recognisable turtles being as old as the earliest dinosaurs. Their survival over that time and adaptation to environments as diverse as the open ocean and the dry deserts are testimony to their evolutionary success. Over those millions of years they have survived mass extinctions that killed off much of the ancient fauna. Their numbers and diversity have fluctuated and climate has played a major role in determining distributions of the turtles. Whatever the changes they have faced there have always been some turtles surviving in every major environment outside the polar regions. Today they face a greater threat than ever before, due mainly to the impacts created by that recently evolved species, man.

The status of the living turtles has still not been fully evaluated. The current Red List of threatened species includes 228 turtle species, of which 141 are considered threatened, but the proportion of species threatened with extinction is probably higher than this indicates. Many of the species are seriously threatened, with populations reduced to critically low levels (such as the single surivivng individual of the Pinta island giant tortoise *Chelonoidis abingdoni*) or by multiple, apparently intractable threats (e.g. all Madagascar species). Many of these are likely to become extinct in the near future, at least in the wild.

What future do turtles have, realistically? This is an important question, but impossible to answer with any certainty. In a changing world the only certain prediction is that ranges, diversity and characteristics will change. This is likely to vary with species, ecosystem and geographical region.

In the seas at least some marine turtles will survive; they have proved themselves great survivors over the past 35 million years, but it is unlikely that they will form the great spectacles they did in the past. The descriptions left by the early European explorers of vast numbers of green turtles around the shores of islands throughout the tropics are already almost impossible to imagine. The nesting arribadas of the ridley turtles are similarly beyond the imagination of most of us, although some people are still lucky enough to witness the few remaining occurrences and the great mass nesting events are still in living memory. It is hard to see how those events could recur. Of all the marine turtle species the leatherback is perhaps most precarious with its diet of jellyfish putting it at risk from the ever increasing amounts of plastic rubbish in the oceans; if this phenomenal animal were to be lost that would indeed be a tragedy.

On land we may see a more varied pattern. Central and South American turtles seem to be relatively resilient; they have adapted to high levels of hunting over thousands of years, but declines in many areas are probably inevitable and the more restricted range species are likely to be lost from the wild as forest habitats in particular shrink. In North America, unpredictable climate change may see the loss of some of the desert tortoises and the freshwater terrapins that are restricted to a few water courses. Similarly the South African restricted range tortoises must be considered highly precarious.

Many of the other African species seem to be quite adaptable, but then many are almost completely unstudied and we still have little idea of the true diversity of species in the Congo. The situation in Madagascar is far clearer and there can be little doubt that there is no realistic future for turtles on that tragic island. Facing the triple threat of dramatically increasing levels of exploitation for food and trade, a rapidly expanding human population and climate change most of Madagascar's tortoises are already on the edge of being beyond salvation.

Unusually, Australia's turtles may be best placed for long-term survival. Their main threat is climate change, and as climate has structured Australian populations perhaps more than those of other geographical regions in the recent past, most species may be able to adapt to that. However, the unpredictable threat may come from trade. At present they are not exploited significantly but with China's insatiable demand for turtles and its proximity it may be only a matter of time before the Australian turtles come under pressure. Trade and human consumption are the main current threats to Asian turtles; some species are already unknown in the wild and the declines in populations are likely to continue. No meaningful consideration has been given to the possible impacts of climate change in that area.

Of all the turtles, giant tortoises have suffered most from human exploitation, with the extinction of some 12 species. With close management of the survivors their future may be more stable, although it is perhaps questionable whether many of these are really wild populations any more. An interesting development in recent years has been the deliberate introduction of Aldabra and radiated tortoises outside of their natural range as 'ecological replacements' for the extinct tortoises of the Mascarenes. Maybe other such assisted movements should be considered to replace missing components of ecosystems, or to enable turtles to change ranges in response to climate change. Many animals and plants are already on the move as their preferred climate zones shift towards the poles; turtles have proven themselves surprisingly good at making such movements in the past, but with the fragmentation of ranges we have caused, few if any are in a position to make these changes any more. These are, and will remain, highly

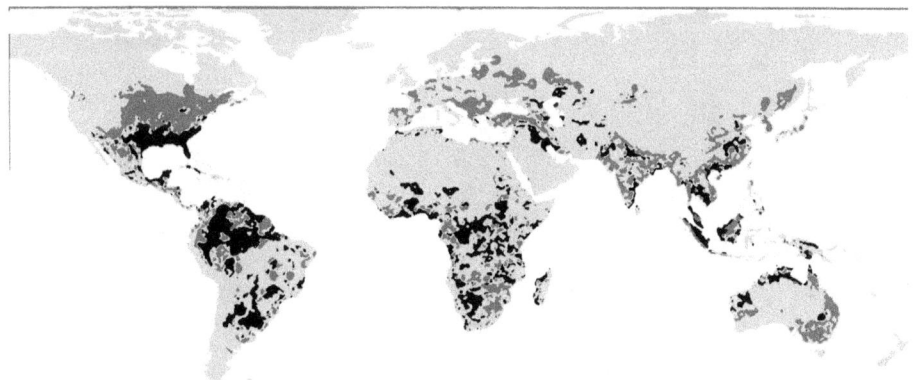

Predicted changes in turtle diversity 100 years from now. Light grey - no change; dark grey - increase in diversity; black - decrease in diversity.

contentious thoughts but consideration needs to be given to the future as never before.

All that is certain is that in the future the distribution and diversity of turtles will be very different. Although they currently face grave threats many species will survive and succeed. For the majority of species that survival will probably not be where those species are today. How they succeed will also probably not be in any way we expect. The 220 million years of turtle history shows one thing clearly, the Chelonia will continue in some form: turtles are the great survivors.

References

Ihlow, F., J. Dambach, J. Engler, M. Flecks, T. Hartmann, S. Nekum, H. Rajaei & D. Roedder. 2012. On the brink of extinction? How climate change may affect global Chelonian species richness. *Global Change Biology*

Index

228